Integrated Renewable Energy for Rural Communities

Planning Guidelines, Technologies and Applications

Integrated Renewable Energy for Rural Communities

Planning Guidelines, Technologies and Applications

NASIR EL BASSAM

International Research Center for Renewable Energy
Sievershausen, Germany

PREBEN MAEGAARD

Danish Center for Renewable Energy
Hurup Thy, Denmark

2004

ELSEVIER

Amsterdam – Boston – Heidelberg – London – New York – Oxford
Paris – San Diego – San Francisco – Singapore – Sydney – Tokyo

ELSEVIER B.V.	ELSEVIER Inc.	ELSEVIER Ltd	ELSEVIER Ltd
Sara Burgerhartstraat 25	525 B Street, Suite 1900	The Boulevard, Langford Lane	84 Theobalds Road
P.O. Box 211, 1000 AE	San Diego, CA 92101-4495	Kidlington, Oxford OX5 1GB	London WC1X 8RR
Amsterdam, The Netherlands	USA	UK	UK

First edition 2004

Library of Congress Cataloging in Publication Data
A catalog record is available from the Library of Congress.

British Library Cataloguing in Publication Data
A catalogue record is available from the British Library.

ISBN: 0-444-51014-1

⊗ The paper used in this publication meets the requirements of ANSI/NISO Z39.48-1992 (Permanence of Paper).
Printed in The Netherlands.

Preface

THROUGH ENLIGHTENMENT TO THE SOLAR AGE

"It is impossible to solve a problem with the same means that caused this problem", Albert Einstein once said.

We have to enlighten our societies about the unique opportunities that Renewable Energies in their immense variety and combinations are offering to mankind. It is, however, impossible to convince people, the millions of users and future energy producers only by developing good technologies and practical instruments. A new strategy without philosophy is like a church without religion. It is essential to discuss not only economic matters, but also values and life benefits.

Renewable Energies are much more than an additional option to the old energy system. This is well illustrated in this book. The Renewable Energies are - when properly integrated - the alternative, the general solution, able to cover all energy needs for lighting, heat, cooking, processing and transportation.

A true realistic picture on the state of the globe shows: To promote renewable energies must become the primary strategy everywhere. Since the problems and dangers of civilization are on Earth and not on the Moon or on Mars, Renewable Energy promotion is more important than space programs. Since dependency on fossil energy makes all nations vulnerable, replacing them is the top security question and more important than new weapon programs.

The narrow and incorrect view about Renewable Energies is the prejudice that they would be an economic burden. The most urgent need is to overcome these mental barriers, including the psychological fear of the consequences of changes for the individuals and society on the whole. No serious economist is in a position to predict the future costs of photovoltaic solar energy conversion or other Renewable Energy options. No one can predict the obvious opportunities of their integration, future applications and their impact on costs, neither the speed of cost reduction as a result of mass production, nor future developments. It is enough to outline the principal opportunity to replace all conventional energies in order to overcome their mythology of being indispensable. Especially in developing countries with their billions of people living in un-served areas only renewable energy can cover urgent needs for improvement. In the industrialised part of the world the merits of Renewable Energies and their integration are becoming more and more obvious.

The prime manufacturers of Renewable Energy products and their system integration are the industries with activities relatively close to solar conversion technologies such as the engine industry, the glass industry, the electrical appliance industry, the electronics industry, the building materials industry, mechanical and plant engineering companies, manufacturers of agricultural equipment, and, not least, agriculture and forestry.

The farmers will become combined food, energy and raw material producers, and they will be ecologically integrated. Ruralisation will lead to new forms of settlement and revitalization of the primary economy with increased local employment and less commuting.

Our farmers will be the oil sheiks of tomorrow. The transport sector, private and public, has an increasing interest to become independent from fossil fuels. It is not only hydrogen, it is especially biofuels that offer an immediate turn to renewable and clean fuels. Just to double the share of Renewable Energy within 10 years is far too modest in order to tackle the real challenges. A lot more dynamic action is needed and this begins in the minds of people. The possibility of completely covering energy demands by means of Renewable Energy sources must be shown for each country and local community, worldwide.   The global energy demand increases faster than the introduction of Renewable Energies. Civilization continues to run into the fossil and atomic energy trap, even with the implementation of the Kyoto-Protocol.

Conventional energies are politically privileged everywhere in the world by large amounts of public money for research and development, by military defence costs, by 300 billion dollars of subsidies annually and by the energy laws tailored to them.   In contrast to this, Renewable Energies are politically discriminated against. Less than a total of 50 billion dollars public money worldwide was spent in the last 20 years to promote Renewable Energy. In order to act as a driving global prime mover for Renewable Energy, the World Council for Renewable Energy, WCRE, adopted in Berlin in 2002 the Action Plan for the Global Proliferation of Renewable Energy with twelve concrete steps for international action. One is to establish an International Renewable Energy Agency (IRENA) in combination with a global network of non-profit technology-transfer centres for the integration of renewable energy in all walks of society for overcoming the double standard of the present institutional system. This includes intergovernmental energy organizations for nuclear and fossil energy such as the International Atomic Energy Agency and the International Energy Agency - but none for Renewable Energies.

To establish IRENA is a world project. It must not happen on a multinational consensus basis because then it will take too long. It is an essential part of the Solar Age. This age will come, sooner or later. If it is later, the problems we face worldwide will be worse. The sooner it comes, the better for society and human race.

The sense of a democratic society is to have a common life based on two values: individual freedom and acceptable social conditions. Individual freedom without touching and restricting the life conditions of other people can only be achieved with Renewable Energy, not by the exhausting old energies. Supporters and promoters of conventional energies may have more influence up to this point, but we, the Renewable Energy promoters have the superior idea for the future.

Dr. Hermann Scheer, General Chairman of World Council for Renewable Energy, WCRE, President of EUROSOLAR

Table of Contents

Index of Figures

Index of Tables

Foreword

Current approaches to energy are non-sustainable and non renewable. Furthermore energy is directly related to the most critical social issues which affect sustainable development: poverty, jobs, income levels, access to social services, gender disparity, population growth, agricultural production, climate change, environment quality, economic and security issues. Without adequate attention to the critical importance of energy to all these aspects, the global social, economic and environmental goals of sustainability cannot be achieved. Indeed the magnitude of change needed is immense, fundamental and directly related to the energy produced and consumed nationally and internationally.

The main goals set in the Kyoto Protocol are, however modest, the mitigation of greenhouse gas emission and thus, if not the reduction, then the stabilization of global warming. Agenda 21, resulting from the United Nation Conferences on the Environment and Development 1992 and Johannesburg 2002, calls for rural energy development. The key challenge is to overcome the lack of commitment and to develop the political will to protect people and the natural resource base. Renewable energy technologies are so well developed, economical and reliable that transition from scarce and polluting fossil fuels to a sustainable energy future should have the highest priority by governments and the world community. Failure to take action will lead to continuing degradation of natural resources, increasing conflicts over scarce resources and widening gaps between rich and poor. We must act while we still have choices. Implementing sustainable energy strategies is one of the most important levers humankind has for creating a sustainable world. More than 2 billion people have no access to modern energy sources, and most of them are living in rural areas. Their share of world population is increasing. Food and fodder availability is very closely related to energy availability. In order to meet these challenges the future energy policies should put more emphasis on the development and deployment of renewable energy resources, forming the foundation of future global energy structure (Brundtland 1987).

The following deserve most grateful thanks: Anja Voges, G. Englert, H. Ahmedsad and Reinhard Hilbert who offered substantial contributions in preparing the manuscript. We would also to thank the members of SREN-Working Group, EUROSOLAR, World Council for Renewable Energies, FAO staff, G. Best and members of worldwide organizations for their contributions and supplying additional information. In particular: J. M. Greef, D. Chiramonti, G. Grassi, H. Scheer, W. Palz, H. Bartelt, F. Hvelplund, N.I. Meyer, P. Gipe, K. Katsube, F. Rutberg, U. Jochimsen, O. Herrera, F. Feitosa, G. Osman, J.Bugge, N. Anso, I. Togola, S.Volchec, G. Best, Li Dajue, M. Trossero, W. Wjewwardene, Wang Mengjie, J. Stolongo, W. A. Kamaruddin, Andre El Bassam, R. Behl, A. Al-Karaghouli, A. Raturi, T.A. Mohamed, D. Christia, H. von Petersdorf, R. Olson, A. Sayigh and J. Fernandez. We are very grateful to Mr. R. Krell, FAO Regional Office for Europe (REU) for all the fruitful discussions and support during the whole period of the preparation of this project. Thanks also to B.F. Brix and to Q. Xi, for their contributions. Most gratitude deserve Mr. Steve B. Smiley and J. Schaellig for reading and correcting the text as well Mrs. Hiltrud El Bassam and Mrs. Jane Kruse for their patience.

Prof. Dr. Nasir El Bassam
Director, International Research Center for Renewable Energy (IFEED), Germany
FAO-SREN, Chairman Working Group 3, Biomass for Energy and the Environment.
Prof. Preben Maegaard Director, Folkecenter for Renewable Energy, Denmark; vicepresident, EUROSOLAR; president, World Wind Energy Association.

1   Introduction

Many of the crises on our planet arise from the desire to secure supplies of raw materials, particularly energy sources, at low prices. The pressure will become even greater as fossil energy feedstock and uranium are depleted. Although some of these resources might last a little longer than predicted, especially if additional reserves are discovered, the main problem of "scarcity" will remain, and this represents the greatest challenge to humanity.

According to Schrempp (2000), a central challenge facing humanity is the question of energy. "Crude oil will begin to get scarce in the next 10 to 15 years. By the end of the current decade – at the very latest – we will face the energy problem severely, in developed as well as in developing countries".

Energy is of crucial importance, since it offers the chance to solve other, subordinate problems at the same time: water, hunger, environmental and climate protection, information, communication, and mobility. The insurance of a sufficient supply of energy -environmentally friendly energy – is one of the biggest challenges facing our planet today.

The world population increases by about one billion every 12-15 years i.e. by the 2050 the world population will amount 8 to 10 billions at the current rate of growth.

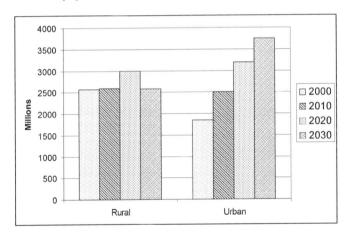

Figure 1:   Urban and rural population projection in developing counties
             (Source: FAO 2000)

Most of the projected population growth is expected to occur in developing countries, and most of that in urban areas. Figure 1 shows demographic projections for developing countries up to 2030. The rural population remains almost stable, whilst there are major increases expected in the urban population. Observations suggest that it is the young people that migrate from rural areas to urban centers, whilst an increasingly aged population is left in the rural communities. These broad projections mask some important regional differences. Regional population projections indicate that a large percentage of Africa's population growth will take place in rural areas, contrary to the trends in Asia and Latin America where rural populations are expected to fall in absolute terms. Given the realities of urban growth and the political and administrative pressure to tackle urban problems, it may become more difficult to keep rural energy development on the agenda.

The greatest challenge of the present century for science and technology is to develop strategies and systems for supplying the increasing world population with food, water and energy. World-wide the food production increases annually around 1% and the population annual growth rate is of 1.7%.

Figure 2 shows annual primary energy consumption per capita in various regions of the world. The data indicate the wide variation between regions, not solely accounted for by climatic differences. Average world annual consumption is around 1.6 toe/capita; in OECD countries the average is around 5 toe/capita and in developing countries it is less than 1 toe/capita.

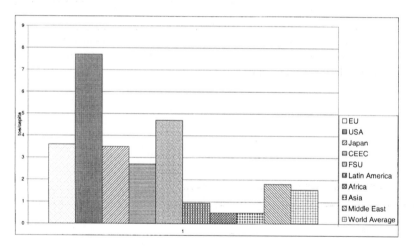

Figure 2:   *Primary annual energy consumption per capita (1990) toe/capita*
              (Source: WEC, 1993)

The last two centuries have seen massive growth in the exploitation and development of energy sources, and the world has gained many benefits from these activities. The magnitude of energy consumed per capita has become one of the indicators of development progress of a country, and as a result, energy issues and policies have been mainly concerned with increasing the supply of energy.

An increase of primary energy consumption per capita especially in Developing Countries is to be expected in the next years and decades (Figure 3).

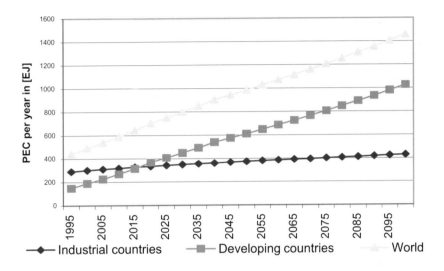

Figure 3: Annual primary energy consumption, Industrial Countries, Developing Countries and World

Also the total primary energy consumption in the Developing Countries will increase much more so that by the year 2020, the total energy consumption in these countries will be higher than in Industrialized Countries (Figure 4).

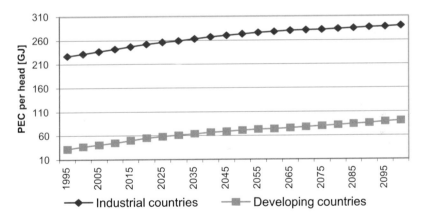

Figure 4: Estimated development of primary energy consumption (PEC) per capita (1995-2100)

Medium and long-term projections show an unavoidable increase in oil prices, which will make biofuels more competitive. Energy crops and biomass in general seem to have, in many instances, much lower external costs than fossil fuels, provided they are grown and processed in an environmentally sound way. It is not helpful to argue the case for bio energy with yesterday's technology without taking into account the innovation possibilities and the prices of today, along with the scarcity, the growing future demand for energy and the environmental damage and risks associated with fossil and nuclear energy feedstock.

The key elements of sustainability in the energy supply are:

- The conscientious development of renewable energy sources, as well as their subsequent utilization.
- Improved energy use efficiency and conservation measures to minimize the loss of primary resources
- The protection of the biosphere, and prevention of more localized forms of pollution.
- The development of adequate political, administrative and financial supports for the implementation and utilization of renewable energy sources.

Economic development is closely correlated with the availability and utilization of modern energy sources. Also, the production and consumption of food is linked to the amount of energy used (Figure 5). In 1990, the per capita consumption of modern energy for 21 African countries was less than 100 kgoe (World Bank, 1992). In many of these countries, daily per capita caloric supply is below 2000 calories. Food production is unlikely to increase without greater access to modern energy (FAO, 1995)

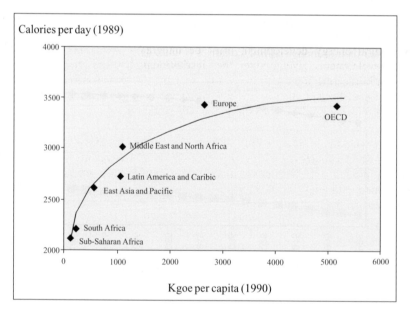

Figure 5:   Energy consumption and calorie supply in different regions of the world
(FAO, 1995)

The implementation of Integrated Energy Farms to produce food and energy for all represents a realistic vision that could be established in different regions of the world (El Bassam 1998).

Renewable energy has the potential to bring power to communities, not only in the literal sense, but by transforming their development prospects. There is tremendous latent demand for small scale, low cost, off-grid solutions to people's varying energy requirements.

People in developing countries understand this only too well. If they were offered new options that would truly meet their needs and engage them in identifying and planning their own provision, then success in providing renewable energy services could become a reality.

Energy is one of the important inputs to empower people provided it is made available to the people in unserved areas on an equitable basis. Therefore access to energy should be treated as the fundamental right for everyone. This is only possible if the end users are made the primary stakeholders in the production, operation and management of the generation of useful energy.

Despite much well intended effort, little progress has been made and a radical new approach is called for based on the following imperatives:

- Rural development in general and rural energy development specifically needs to be given higher priority by policy makers.
- Rural energy development must be decentralized and local resources managed by rural people.
- Rural energy development must be integrated with other aspects of rural development, overcoming the institutional barriers between agriculture infrastructure and education as well as in the social and political spheres.
- The problem of non-committing policies on the government level, different priorities in each country should be addressed.

The productivity and health of a third of humanity are diminished by a reliance on traditional fuels and technologies, with women and children suffering most. Current methods of energy production, distribution and use worldwide are major contributors to environmental problems including global warming and ecosystem degradation at the local, regional and global levels.

2 Overview of Energy Requirements for Rural Communities

Energy supply in rural communities has to meet the needs of the people and to ensure economic and social development. In order to generate adequate energy, it is necessary to determine the most appropriate and affordable technologies, equipments and facilities. Basic elements of the required needs of such communities are illustrated in Figure 6. The technologies and their applications are described in the next chapters.

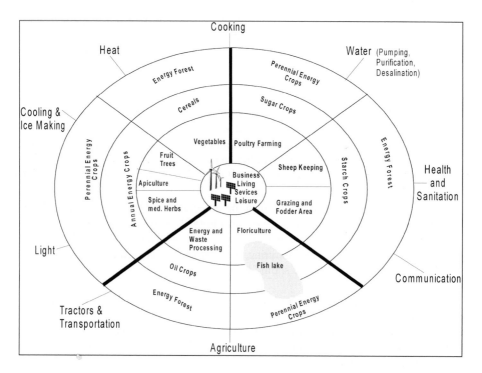

Figure 6:   Basic Elements and Needs of Integrated Renewable Energy Community (El Bassam 2000)

2.1   Heat

Heat can be generated from biomass or solar thermal to create both high temperature steam and low temperature heat for: space heating, domestic and industrial hot water, pool heating, desalinsation, cooking and crop drying.

2.2   Electric Power

Solar PV, Solar thermal, biomass, wind, micro-hydro.

2.3   Water

(Drinking and Irrigation)

Water is an essential resource for which there can be no substitute. Renewable energy can play a major role in supplying water in remote areas. Several systems could be adopted for this purpose:

- Solar distillation.
- Renewable energy operated desalination units.
- Solar, wind and biomass operated water pumping and distribution systems.

2.4   Lighting

In order to improve living standards and encourage the spread of education in rural areas, a supply of electricity is vital. Several systems could be adopted to generate electricity for this purpose:

- Solar systems (photovoltaic and solar thermal)
- Wind energy systems
- Biomass and biogas systems (engines, fuel cells, Stirling)

2.5   Cooking

Women in rural communities spend long hours collecting fire wood and preparing food. There are other methods which are more efficient, healthy and environmentally benign. Among them are:

- Solar cookers and ovens
- Biogas cooking systems
- Improved biomass stoves using briquettes and pellets
- Plant oil and ethanol cookers

2.6   Health and Sanitation

To improve serious health problems among villagers, solar energy from photovoltaic, wind and biogas could be used to operate:

- Refrigerators for vaccine and medicine storage
- Sterilisers for clinical items
- Waste water treatment units
- Ice making

2.7   Communications

Communication systems are essential for rural development. The availability of these systems has a great impact on people's lives and can advance their development process more rapidly. Electricity can be generated from any renewable energy source to operate the basic communication needs such as: radio, television, weather information systems, and mobile telephone.

2.8   Transportation

Improved transportation in rural areas and villages has a positive effect on the economic situation as well as the social relation between the people of these areas. Several methods could be adapted for this purpose.

- Solar electric vehicles
- Ethanol, plant oil fuel, hydrogen vehicles (engines, fuel cells)
- Traction animals.

2.9   Agriculture

In rural areas agriculture represents a major energy end use. Mechanisation using renewable sources of energy can reduce the time spent in labour intensive processes, freeing time for other income producing activities. Renewable sources of energy can be applied to:

- Soil preparation and harvesting:
- Husking and milling of grain
- Crop drying and preservation
- Textile processing

2.10   Basic and Extended Needs

Power supply systems from renewable sources in off-grid operation should be robust, inexpensive and reliable. Most importantly they need to have a modular structure so they can be extended later. Photovoltaic power supply systems in off-grid operation supply power to small consumers (3-30 kW) far from the public utility grid. An essential component of a modular supply system is the battery inverter such as the Sunny Island with a nominal power of 3.3 kW each.

The advanced battery inverter Sunny Island is the grid master and the central component of a modular supply system and enables small scale island utilities for remote areas. An island grid is easy to plan and to install and allows a very flexible operation.
The device includes an intelligent control which is able to supply different consumers and feed power from different generators. Such generators are e.g. PV-String Inverters for grid supply, small wind energy plants or diesel units, making the battery inverter operate on all four quadrants. The system management handles battery control, enables limited load management and provides communication interfaces for optional system management units. The required operating modes and parallel switching of current converters can be realized. A battery inverter and a lead acid battery can establish a simple single-phase island grid.

2.10.1   Typical electricity needs in rural areas

In order to be able to size the power supply properly, the peak power, the daily energy consumption and the annual energy growth should be estimated. This is necessary in order to size both the generating plant and the conductor used in the distribution system. The sustainability of the system will not be guaranteed if the capacity of the system is too small and leads to consumer dissatisfaction. On the other side, alternatively, too much capacity would mean additional investment costs and possibly too high tariffs.

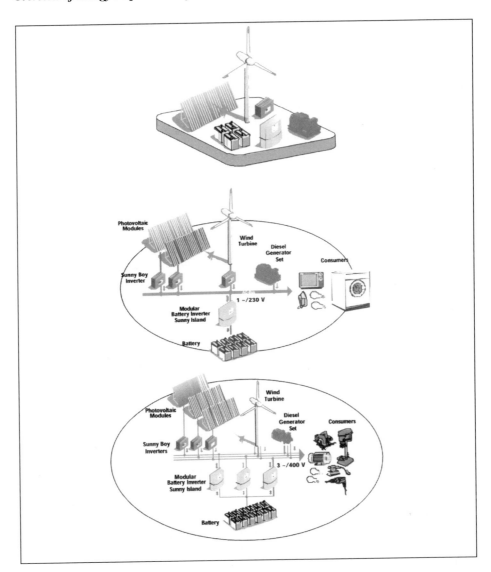

Figure 7:   Main components of a PV Hybrid System, a single-phase island grid, and three-phase island grid

Making load projections that reflect reality is frequently a difficult task to accomplish, especially for prospective consumers who have little experience with electrification. The more reliable approach to assess demand is to survey households in adjoining, already-electrified areas or in a region with similar economic activities, demographics characteristics etc. This would assess the average initial loads per household in these areas as well as their historical load growth. The already-electrified regions that would be surveyed should preferably have a similar type of service as that being proposed in the new community, such as 24-hour power or electricity for 4 hours each evening.

Any projections of load and load growth in an area to be electrified using information gathered from already electrified regions should also consider such factors as the difference in the level of disposable income in the two areas, the presence of raw materials or industry, the potential for tourism and access to outside markets for goods which might be grown or produced locally.

Typical appliances in remote areas are:
- Lighting
- Television and radio
- Refrigerator
- Other appliances like irons, hot plates, cookers, hair dryers etc.

Also the electricity is often needed for central services like:
- Water pumps
- Water treatment
- Electrification of a school or acute day ward

Regarding the needs for electrification it is possible to divide this in three different classes:
- Basic needs,
- Extended needs and
- Normal needs

For people who are accustomed to the use of electricity. The following table gives an overview to the daily electricity demand for these three classes:

	Basic Needs	Extended Needs	Normal Needs
User Appliances			
Lighting	3 x 11W x 3h	4 x 15W x 4h	4 x 15W x 4h
TV / Radio	30W x 4h	30W x 5h	60W x 5h
Refrigerator		10W x 24h	30W x 24h
Others	100 Wh	300 Wh	1,500 Wh
Central Services for 50 Households			
Water Pumps	3,000 Wh	6,000 Wh	10,000 Wh
Water Treatment	1,500 Wh	3,000 Wh	5,000 Wh
Others	1.500 Wh	1,.000 Wh	20,000 Wh
Daily Consumption / Household	440 Wh	1,310 Wh	3,360 Wh
Daily Consumption / Village [2]	22 kWh	65.5 kWh	168,000 kWh
Average Power	0.92 kW	2.73 kW	7 kW
Peak Power	Ca. 4 kW	11 kW	21 kW

Due to the high cost of electricity generation, it is very important to choose the most efficient appliances. The basis for all assumptions in the table is the use of such appliances. Otherwise the energy consumption will increase significantly. The peak power is typically three to four times higher than the average power.

Example for electricity supply of small villages with approximately 50 inhabitants

In power generation, PV plants, wind or hydro-electronic power plants can be combined. Normally an additional electric generator such as a diesel generator can make the supply more reliable. The controls allow for a power increase by switching up to three battery inverters in parallel on one phase. In this example every third household has a refrigerator and every village consists of 50 households.

Version 1: Single-phase island grid
Combination of power generation with parallel operation of inverters.

Version 2: Three-phase island grid
If there is the necessity to connect three phase consumers the design of the island is flexible and extendable. The smallest three-phase system has a nominal output power of 10 kW and consists of 3 Sunny Island inverters. 3-phase systems also simplify the connection of larger diesel sets and wind energy systems. These are mostly equipped with 3-phase generators.

Version 3: Three-phase island grid and parallel operation of the sunny island inverter
Several Sunny Islands can be combined to establish a three-phase system up to 30 kW

The System Solution for Island Grids:
- Simple design of island grids due to connection of all components on the AC side
- Reliable and safe power supply with utility quality in remote areas
- Easy integration of photovoltaic plants and wind energy or diesel generators
- Power supply for single houses or even small villages
- Extendable design (1 – or 3 – phase combinations, parallel operation)
- Optimal battery life

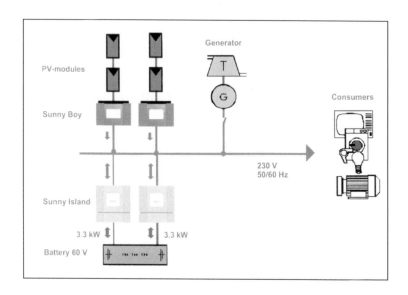

Figure 8:   *Integration of PV-Plants and Diesel set and parallel operation of Sunny Island inverters*

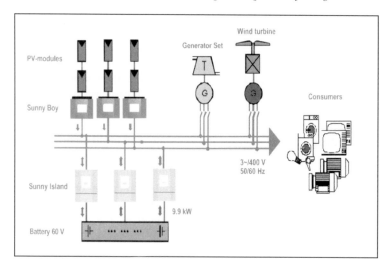

Figure 9:   *Integration of PV-Plants, Diesel set and parallel operation of Sunny Island inverters*

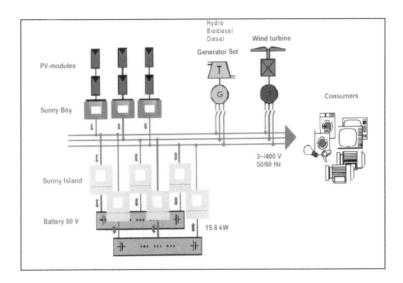

Figure 10: *The parallel operation of Sunny Island inverters suitable for high performance.*

3 Calculating Energy and Food Production Potential and Requirements

3.1 Modeling approaches

The modeling procedure should include the identification and determination of the following parameters

1. Site conditions:
   - climate: temperature, amount and distribution of precipitation, sunshine duration, wind velocity (annual mean)
   - soil conditions, irrigation possibilities etc.
   - factors of production: capital, machines, building, agricultural area.
2. Energy requirement per year for food production for households
3. Basic food requirement per person per year
4. Number of the energy consumers (persons and households)
5. Site energy potential: solar energy, wind energy and biomass
6. Preparation of a master production schedule for food and energy production
7. Selection and installation of suitable technical tools using the renewable energy resources of the site
8. Energy production and use management
9. Environmental impact
10. Social and economic impact

For elaboration, development and establishment of the Integrated Energy Farm (IEF), two scenarios have been considered:

a) **Scenario 1** (Flow chart Figure 11)
   Initial conditions: The farm size (Ftn) is known, i.e. an already existing farm, or the agricultural area is limited.
   Objectives: The available area (Ftn) should be managed to achieve a high degree of self-sufficiency for a maximum number of people with basic food (Nx) and energy
b) **Scenario 2:** (Flow chart Figure 12)
   Initial conditions: The farm size (Ftx) is variable, i.e. the size could be adapted according to needs
   Objectives: High degree of self-sufficiency for a determined number of people with basic food and energy should be achieved.

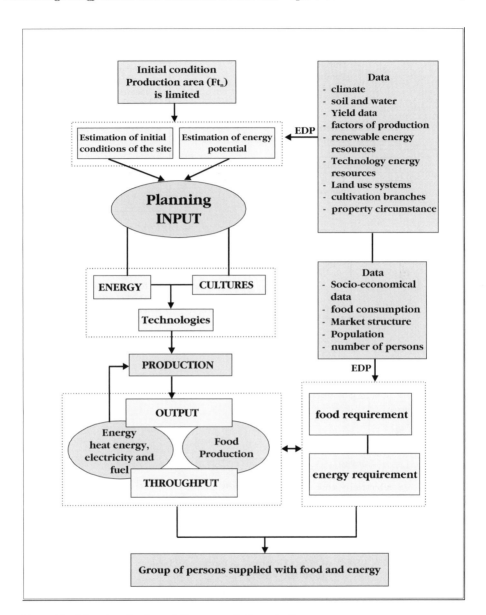

Figure 11: Flow chart for the modelling approach (Scenario 1)

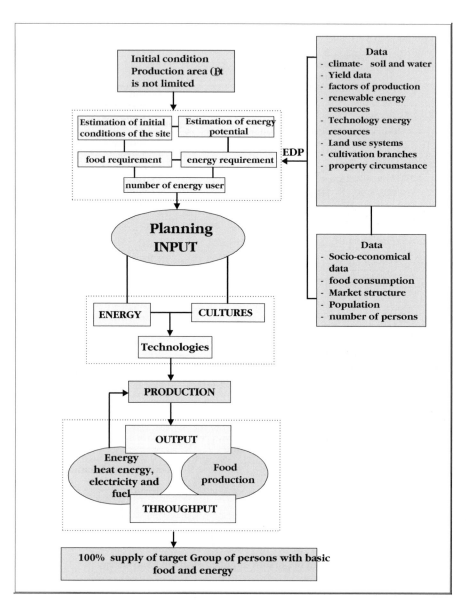

Figure 12: Flow chart for the modelling approach (scenario 2)

3.2　Data acquisition

The following data should be identified in details for the planning, modelling and implementation of an integrated energy farm:

Table 1:　*Data acquisition (Overview)*

	To be measured	To be calculated	To be recorded
External Data:			
a)　*Climate data:*			
• precipitation and distribution,	X		X
• temperature,	X		X
• annual temperature variations			X
b)　*Socio-economic data:*			
• size of the population			X
• property structures			X
• age structure of the population			X
• education			X
• economy data, number of trade companies, industry			X
• employment situation			X
• land use: agriculture and forestry			X
• land division: crop production, animal breeding, fruit cultivation, vegetable, pisciculture, forestry area etc.			X
• number of households			X
• food consumption			X
Farm Internal Data:			
a)　*General*			
• farm size	X		X
• land use system: crop production, fodder production, forestry area		X	X
• properties of the soil			X
• infrastructure			X
• mechanisation degree			X
• soil condition and quality			X
• availability of manpower			X

	To be measured	To be calculated	To be recorded
Farm Internal Data:			
a) **Production branches**			
• crop production: cereals, oil plants, root plants, fodder production: grassland, field fodder culti-vation		X X	X X
• market gardening			
• animal husbandry: dairy cattle, pigs, sheep's, chicken farming etc.		X X	X X
• fruit trees		X	X
• pisciculture etc.		X	X
b) **External and internal inputs**			
• organic and mineral fertilisers		X	X
• pesticides		X	X
• fuel		X	X
• machines		X	X
• capital		X	X
c) **Yield data**			
• crop production (type of cultures)		X	X
• animal husbandry		X	X
• other branches: gardening, pisciculture etc.		X	X
d) **Marketing and procurement**			
• methods of marketing			X
• market products			X
• proximity to market			X
• affiliation to production / marketing / procure-ment organisations and / or associations			
			X

	To be measured	To be calculated	To be recorded
Energy requirement (fuel, electricity and heat)			
a) Agricultural sector			
• soil preparation		X	
• cultivation		X	
• harvest		X	
• transport and storage		X	
• cooling		X	
• drying		X	
• transformation		X	
• lighting		X	
• pumping		X	
b) Administration + household			
• space heating and cooling		X	
• lighting		X	
• cooking		X	
• warm water supply		X	
• cold storage		X	
• communication, radio, TV etc.		X	
c) Power demand (per consumer)			
• peak demand	X		
• daily average	X		
• annual sum		X	
Energy Supply			
• regional distribution of primary energy supply			X
a) Solar energy			
• number of sunny days			X
• degree of cloudiness			X
• global radiation on surface with optimum inclination			
• value during the course of the day	X		
• daily sum		X	
• annual sum		X	

	To be measured	To be calculated	To be recorded
Energy requirement (fuel, electricity and heat)			
• global radiation and diffuse radiation on horizontal surface			
• value during the course of the day	X		
• daily sum		X	
• annual sum		X	
b) Terrain			
• supply of cooling water available (yes/no)			X
• cooling water temperature			X
• max. quantity of cooling water available		X	
• transport routes		X	
• survey of terrain			X
• shading (houses, trees)			X
• reflections (bodies of water)	X		
• salt content of air (occasional measurements)			X
c) Wind energy			
• wind intensity	X		
• wind speed	X		
• wind direction	X		
• other limiting factors of installation plants	X		
d) Biomass			
type and quantity of biomass available			
• wood and wood residues		X	
• straw		X	
• bio wastes		X	
• other bio materials		X	
• transport routes			X
• distance from field to power house			X
• type of oil plants			X
• surface available for cultivation		X	
• yield data		X	

	To be measured	To be calculated	To be recorded
Energy requirement (fuel, electricity and heat)			
possibilities of cultivation of "energy plants"			
• type of cultures			X
• surface requirement		X	
• surface availability			X
• yield data		X	
• equipment requirement		X	
• type of exploitation		X	
• transformation		X	
• transport			X
• storage			X
e) Other important data			
• nearest mechanical workshop			X
• availability of spare parts			X
• distance to spare parts depot			X
• regional distribution of consumers (map)			X
• marketing possibilities of energy			X
• existence of other producers of bio energy raw materials			X

3.3 Determination of energy and food requirements

3.3.1 Agricultural activities

Agriculture is itself an energy conversion process, namely the conversion of solar energy through photosynthesis to food energy for humans and feed for animals. Primitive agriculture involved little more than scattering seeds on the land and accepting the scanty yields that resulted. Modern agriculture requires an energy input at all stages of agricultural production such as direct use of energy in farm machinery, water management, irrigation, cultivation and harvesting. Post-harvest energy use includes energy for food processing, storage, and transport to markets. In addition, there are many indirect or sequestered energy inputs used in agriculture in the form of mineral fertilizers and chemical pesticides, insecticides and herbicides.

Whilst industrialized countries have benefited from these advances in energy availability for agriculture, developing countries have not been so fortunate. "Energizing" the food production chain has been an essential feature of agricultural development throughout recent history and is a prime factor in helping to achieve food security. Developing countries have lagged behind industrialized countries in modernizing their energy inputs to agriculture.

Agriculture accounts for only a small proportion of total final external commercial energy demand in both industrialized and developing countries. In the OECD countries, for example, around 3-5% of commercial energy consumption is used directly in the agricultural sector. In developing countries, estimates are more difficult to find, but the equivalent figure is likely to be similar – in the range of 4-8% of total final commercial energy use.

The data for non-renewable energy use in agriculture also excludes the energy required for food processing and transport by agro-industries. Estimates of these activities range up to twice the energy reported solely in agriculture. Definitive data does not exist for many of these stages, and this is particularly problematic in analysing developing country energy statistics. In addition, the data conceals how effective these energy inputs are in improving agricultural productivity. It is the relationships between the amounts and quality of the direct energy inputs to agriculture and the resulting productive output that are of most interest.

Looking more closely at energy use in specific crops, comparisons of commercial energy use in agriculture for cereal production in different regions of the world are listed in Table 2. The relationship between commercial energy input and cereal output per hectare for the main world regions is also shown in Figure 13. These data, whilst relatively old, indicate that developing countries use less than half the commercial energy input (whether in terms of energy per hectare or arable land or energy per ton of cereal) compared with industrialized countries. However, this is not to say that developing countries are necessary more efficient in their use of energy for agricultural production.

Table 2:   Commercial energy use and cereal output (1982)
(Source: Stout, 1990)

Region	Energy per hectare of arable land (kgoe/ha)	Energy per ton of cereal (kgoe/t)	Energy per agri-cultural worker (kgoe/person)
Africa	18	20	26
Latin America	64	32	286
Far East	77	43	72
Near East	120	80	285
All developing countries average	96	48	99
All industrialized countries average	312	116	3294
World average	**195**	**85**	**344**

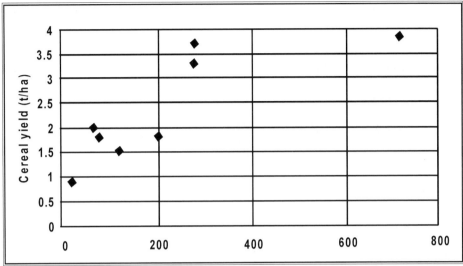

Figure 13:   Cereal yield and energy input per hectare for the main world regions

A comparison between the commercial energy required for rice and maize production by modern methods in the United States, and transitional and traditional methods used in the Philippines and in Mexico is shown in Table 3. These data show that the modern methods give greater productive yields and are much more energy-intensive than transitional and traditional methods. These methods include the use of fertilizer and other chemical inputs, more extensive irrigation and mechanized equipment.

Table 3:   Rice and maize production by modern, transitional and traditional methods
(Source: Stout 1990)

	Rice production			Maize production	
	Modern (United States)	Tran-sitional (Philippines)	Traditional (Philippines)	Modern (United States)	Traditional (Mexico)
Energy input (MJ/ha)	64,885	6,386	170	30,034	170
Productive yield (kg/ha)	5,800	2,700	1,250	5,083	950
Energy input yield (MJ/ha)	11.19	2.37	0.14	5.91	0.18

A further illustration of the relationship between energy intensity and agricultural productivity is given in Figure 14. This presents a case study of energy use in durum wheat production in Tunisia.

The study showed that farms with the highest energy input per hectare had the highest production and lowest input per ton of production.

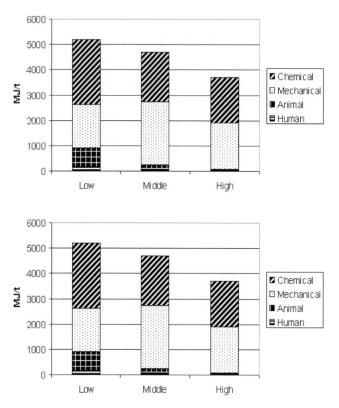

Figure 14:   Energy use in Durum Wheat Production in Tunisia
(Source: Myers, 1983)

The energy data for production **(Ep)** depend first on the size of the farm **(Sq)**, the mechanization degree and the production activities. There is the requirement of fuel **(Ep1)** and of electricity **(Ep2)**.

The fuel is required for activities of soil preparation and cultivation, as well as for harvesting and transportation.

Fuel requirement **(Ep1)** is calculated as follows:

$$Ep1 \ (l/y) = \text{working duration/ha} \times \text{area (Sq1)} \times \text{fuel consumption /h/Machine}$$

$$Ep1 \ (\text{Litre/year}) \ = \sum_{1}^{n} Sh/ha \times Sq1 \times al \qquad (1)$$

Sh	=	working duration per machine and per ha
Sq1	=	size of the field
Al	=	fuel consumption per machine and per hour
1...n	=	different cultivation branches
ha	=	hectare

The electricity requirement depends on the mechanization degree. Electricity is required especially in animal production but also for the storage, cooling and drying of crops. The energy requirement for production **(Ep2)** is the sum of all usage factors of all electric instruments and machines to be used on the farm.

$$Ep2 \ (MWh/y) \ = \sum_{1}^{n} h/y \times al_1 \qquad (2)$$

h/a	=	Working duration per machine and per year
al_1	=	Power requirement per machine and per hour
1...n	=	different cultivation branches

Considering formula (1) and (2) the total energy requirement for agricultural production is to be calculated as follows:

$$Ep = Ep1 + Ep2 + \sum_{1}^{n} \alpha$$

α	=	losses in kWh/a/machine

3.3.2   Households

The energy requirement for the households **(Eh)** is divided into heat energy (for heating, cooling and hot water preparation) and power requirement for light, electrical appliances and for cooking. The requirement data are different from region to region depending on climatic conditions and should be calculated considering the specific site conditions.

a)   Heat energy

The requirement data for heating and cooling load **(Eh1/household)** depends in addition to the site and environmental conditions and also on the construction type of buildings (full insulation, half insulation). Heating and cooling loads for the buildings are calculated as follows:

$$\text{Eh1 (MWh/y)} = \gamma \ (W/m^2) \ x \ h/y \ x \ F(m^2) \qquad (3)$$

$\gamma \ (W/m^2)$	=	energy requirement per reference area
h/y	=	year requirement of energy in full load hour
$F(m^2)$	=	Energy reference area

Hot water requirements are indicated considering different use possibilities in litre/day/person. The total heat energy requirement for hot water with a temperature of 60 °C is determined as follows:

$$Q \ (kJ/y) \ = \ m \ . \ (t2 - t1) \ . \ 4.19 \ KJ \ / \ K \ . \ 365$$
$$Q \ (kJ/y) \ . \ 0.000278 = Q \ (kWh/y)$$

Q	=	heat energy requirement
m	=	hot water consumption per day (l)
t1	=	cold water temperature (10 - 20 °C)
t2	=	desired temperature (60 °C)

b)   Electricity

The total requirement (Eh2/Household) results from the usage data of all electrical devices existing in a household for different uses: light, communication, cooking and cooling.

$$\text{Eh2 (MWh/y)} = \sum_{1}^{n} \beta \ (kW) \ x \ h + \alpha \ (kWh) \qquad (4)$$

$\beta \ (kW)$	=	connected load of the device
h	=	full load hours in the year
α	=	losses in kWh

Considering formula (3) and (4) the total energy requirement of the household is calculated as follows:

$$Eh\ (MWh/y) = \{Eh1(MWh/y) + Eh2(MWh/y)\} \times n$$

n = number of households

3.3.3   Food requirement

On a broad regional basis, there appears to be a correlation between high per capita modern energy consumption and food production. Figure 15 shows data for daily food intake per capita and the annual commercial energy consumption per capita in seven world regions (FAO, 1995)[*].

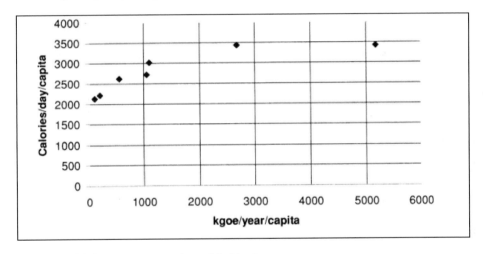

Figure 15:   Modern energy consumption and food intake

Whilst broad data on a regional basis conceal many differences between countries, crop types and urban and rural areas, the correlation is strong in developing countries, where higher inputs of modern energy can be assumed to have a positive impact on agricultural output and food production level. The correlation is less strong in industrialized regions where food production is near or above required levels and changes in production levels may reflect changes in diet and food fashion rather than any advantages gained from an increased supply of modern energy.

Basic food requirement (person/year) is divided into requirement data for carbohydrates, proteins, vitamins and fats. They are different from country to country. For the realization of an integrated energy farm and also for the preparation of a land

[*] The regions are: Sub-Saharan Africa, South Asia, East Asia/Pacific, Latin America / Caribbean, Middle East / North Africa, Europe, OECD

use plan, the regional and national consumer data should be used. In the table 4, the basic foods and their resources from agricultural production are performed.

The data for the food consumption are normally known and must be used while planning. Which kind of food will be finally produced on the farm depends on the climatic and soil conditions of the site.

Table 4:   Basic foods and their sources

Food components	Sources	Indicators
Carbohydrates	Cereals	kg/person/year
	Other crops	kg/person/year
Protein	Animal	number of animals/farm
	Plant	kg/person/year
Vitamin	Fruits	kg/person/year
	Vegetables	kg/person/year
Fats	Oil plants	kg/person/year
	Animal	number of animals/farm

For the determination of the production area for each agricultural branch the following calculation formula can be used.

$$\mathbf{Ar_{1...n}} \quad = \quad (\mathbf{Y_{1...n}}) \, / \, (\mathbf{F_{1...n}})$$

Ar	**=**	**space requirement in ha**
Y	**=**	**yield / area**
F	**=**	**requirement/head/year**
1...n	**=**	**different basic food**

3.4   Energy potential analysis

3.4.1   Solar energy

The total solar radiation that strikes the earth surface amounts to 1018 kWh/a, which is many times greater than the present global energy demand.

In the case of vertical incidence (solar altitude: 90°), the radiation intensity of global radiation can reach a level of 1100 w/m^2.
The daily sum of global radiation (horizontal surface) on a sunny day in the vicinity of the equator is estimated to be 6 – 8 kWh/m^2/ d.

It is possible to estimate the proportion of global radiation represented by diffuse radiation (important for the utilization of solar collectors). It is a function of the solar altitude and the degree of cloudiness and ranges from 10 – 85%.

The available data regarding regional solar radiation come from measurements on horizontal surfaces. But to increase their efficiency, the solar collectors are usually mounted at tilt angle. The solar radiation is divided into two components: the direct radiation and the diffuse radiation. The conversion of the direct radiation is relatively simple. However, specific assumptions must be made for conversion of the diffuse component, because this radiation component is extremely dependent on site conditions and on technical facilities. Generally three estimating procedures are possible.

1. The first assumes that the sky radiation becomes the predominant part of direct solar proximity. This can be possible only on clear days. The total radiation falling on a tilt surface GG,g is then:

$$G_{G,g} = RG_{G,h} \; (W/m2)$$

R = ratio of the direct radiation on a tilt surface ($G_{o,g} = G_o \cos \varphi$) to the radiation value on a horizontal surface

2. The second assumes that the sky radiation is distributed uniformly via the entire sky vault. This is an approximation for cloudy days. The entire radiation falling on tilt surface results to:

$$G_{G,g} = RG_{D,h} + RG_{H,h} \; (W/m^2)$$

$G_{D,h}$ = direct radiation on horizontal surface
$G_{H,h}$ = sky radiation on horizontal surface

3. The third procedure represents a middle way between the two extremes. One assumes here that because of the inclination of the absorber, it sees only a part of the sky vault (indeed ½ (1+cos n), but it receives an additional diffuse part of radiation in the form of ground reflection from the collector environment. The total radiation falling on a tilt surface consists of three parts, direct component, the diffuse component and the reflected part:

$$G_{G,g} = RG_{D,h} + [\; ½ \; (1+\cos n)] \; G_{H,h} + \sigma_B \; G_{G,h} \; (W/m^2)$$

n = inclination angle of the receiving face (degree)
σ_B = reflection coefficient of the surrounding ground

With the procedures performed above the solar energy potential of a site can be estimated. Extensive local measurements are necessary in order to evaluate the utilization potential of the solar radiation energy in a particular region.

Solar installation sites must be carefully selected. The primary energy supply and the presumed energy demand are the decisive factors in determining the economic feasibility of a particular site. Local measurements should include the following quantities:

- global radiation G
- direct solar radiation S
- diffuse sky radiation H
- number of hours of sunshine SD
- degree of cloudiness N
- air temperature T_A
- wind direction D
- wind intensity F

If possible, the measurement should be conducted over a relatively long period (several years). The transformation of the sun energy to electricity by photovoltaic panels and heat energy by solar thermal collectors depends on the type and model of collectors. Therefore, the efficiency of the chosen collector should be considered for the calculation of site solar energy potential.

Figure 16:   Applications of solar energy
Photovoltaic and   micro windmill power submerged pumps by Grundfos (left); 6000 sqm
thermal solar energy deliver district heating to the town of Marstal, Aeroe (middle); solar
shading with photovoltaic panels improves room comfort and produces electricity (right).

3.4.2   *Exploitation of solar energy*

Today, solar energy is being utilized in many ways at various scales. On a small scale, it is used at the households' level in goods such as watches, cookers and heaters. The medium-scale uses, such as in solar architecture houses, include water heating and irrigation. At the community level, it can be used for water pumping, water desalination, purification and rural electrification. On an industrial scale, solar energy is used for power generation, detoxification, municipal water heating and telecommunication. In general, there are actually two basic ways to use solar energy.

3.4.3   *Solar thermal system*

While heat from the sun (over 300 °C) is utilized on a large scale in electricity generation, it can also be used in small to medium scale heating, cooling, cooking and drying equipment. Solar thermo-electric technologies utilize energy from the sun in the form of heat to generate electricity. The sun evaporates a fluid from which heat transfer systems may be used to operate an engine that drives a power conversion system. In a solar thermo-electric system, sunlight is concentrated with mirrors or lenses to attain a high temperature sufficient for power generation. Parabolic trough systems, central-receiver systems, parabolic dish systems and solar ponds are among those used.

Figure 17:   Testing of high temperature solar collectors, Sandia Laboratories, USA
Trough with collector pipes in focal point (left); solar tracking parabols with boilers (right)

The basic components of a solar thermo-electric system are a collector system (which is the panels that collect the solar radiation), a receiver system, a transport storage system (mainly in the form of fluid that transfers the heat between the systems) and lastly a power conversion system converting energy from one form to another.

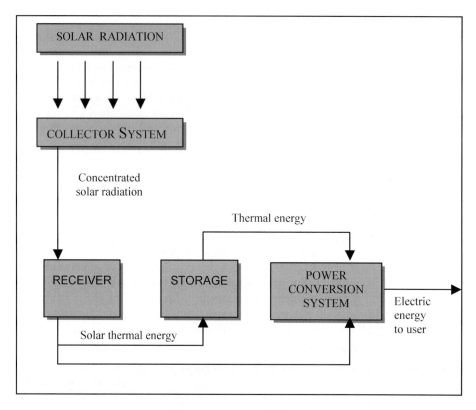

Figure 18:   Main components of a solar thermo-electric system
Source: Adapted from De Laquil et al., 1993

Solar water heaters are relatively simple solar thermal applications that transform solar radiation into heat that is used to warm water for heating, washing, cooking and cleaning. Solar water heaters consist of glass-covered collectors with a dark coloured or especially coated absorber panel inside.   Water (used as the heat transfer fluid) is warmed by the sun and can be stored in insulated tanks for later use. There are two main components of a typical solar water heating system: the flat plate solar collector and the hot water storage tank. Flat plat collectors absorb the solar radiation and conduct the heat to water that circulates through the collector in pipes.

Figure 19: Solar thermal warm water supply: Test station with fifteen different types at Folkecenter Denmark (top); MEGASUN integrated, self circulating solar heater (left): high-efficiency vacuum tubes mounted on roof (right).

Another solar thermal application is the solar drier. There are often two stages of the process; first, solar radiation is captured and used to heat air; then comes the actual drying during which heated air moves through, warms and extracts moisture from the product. Drying takes place in a large box called the drying chamber. Air is either heated in a flat plate collector or directly via a window in the drying chamber.

All these components of the solar thermal system could be used on a farm and contribute by minimizing the costs of energy supply.

Table 5:   Various applications of solar thermal systems in an agricultural farm

Temperature range	Applications
Low grade thermal energy ‹ 100 °C	Water heating, air heating, space heating, space cooling, communication etc.
Medium grade thermal energy 100-300 °C	Cooking, drying, pumping, Irrigation, Water desalination

3.4.4   Solar photovoltaic

Solar cells are produced from wafers of silicon (a form of pure sand) which is chemically treated and then arranged in parallel or in series in a module /panel. It can also be a film coating that is applied to a glass plate. The photovoltaic effect occurs when light falls on an active photovoltaic surface. This energy penetrates the cell near the junction between the p-type and n-type silicon. The semi-conductor dislodges an electron, leaving behind a 'hole'. The electrons generated in this process of electron-hole pair formation tend to migrate to the n- region in the front contacts and can flow into an external circuit.

The main components of a PV system (figures 16, 17) include a module, a battery, a battery control unit/charge controller, a DC-AC inverter (where necessary) and the load (appliances).

Important applications of PV-systems in agriculture include:

- Electrification (lighting for buildings, power supply to remote locations)
- Solar pumps for water pumping
- Household and office appliances (ventilation, air conditioners, computers, emergency power, battery chargers etc.)
- Communication (PV-powered remote radio telephones or repeaters)
- Solar desalination

*Figure 20:   Solar power finds many applications: As roof of IKEA headquarters in Sweden(left); in reno-
vated facade in Vienna (middle); and lighting of market place in Niamala village, Mali
(right).*

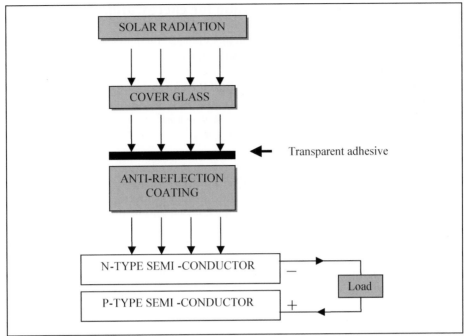

Figure 21: Flowchart of a solar cell

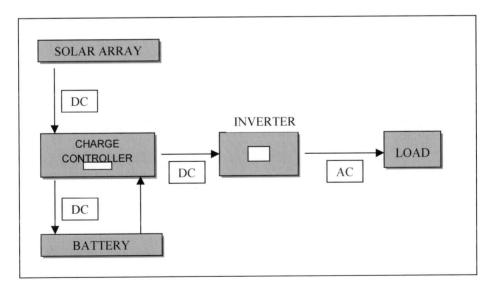

Figure 22: Flowchart of main structure of a PV system

3.5   Data collection and processing for energy utilization

The following basic data on solar radiation and climate conditions are essential for the development of preliminary designs for the solar plants:

- average global radiation density Q_i *(kW/m²)*
- average direct radiation density Q_1 *(kW/m²)*
- average duration of sunshine   *S (h/d or h/a)*
- average ambient temperature T_u *(°C)*

The local global radiation density Q_i , the duration of sunshine, the temperature T_u and indicators for the direct solar radiation Q_1, should be obtained from the records of national measuring stations. The annual curves of both the average daily and monthly values, as well as the average annual values, are relevant information.

In addition, at the site of a projected solar installation, measurement should be taken of the global radiation level, the direct solar radiation level, the ambient temperature and the atmospheric humidity. From the humidity data general conclusions can be drawn as to the climatic conditions as well as the variations exhibited by Q_i, Q_1, T_u and S.

3.5.1   Water and space heating:

For water and space heating, the potential consumer must first define the energy requirement Qv. It is calculated as follows:

$$Q_v = m_v \cdot C_1 \cdot (T_2 - T_1) \ (kJ/h) \quad (1)$$

Where m_v = quantity of water to be heated per hour (kg/h), C_1 = specific heat of water = 4.18 (kJ /Kg deg. [°C]), T_1, T_2 = initial temperature of the cold water, final temperature of the warm water (°C).

The energy supply Q_s of a solar installation can be calculated as follows:

$$Q_s = \vartheta \cdot Q_i \cdot A \cdot \cos \alpha \ (kJ/h) \quad (2)$$

where ϑ = overall efficiency of the solar plant; Qi = average global radiation density (kJ/m2 h); A = collector area of the solar installation (m2); α = angle between a line perpendicular to the surface and the direction of radiation.
By combining equations (1) and (2) the required collector area A can be obtained as follows:

$$A = m_v \cdot c_1 \ / \ \vartheta \cdot Q_i \cdot \cos \alpha \ (T_2 - T_1) \ (m^2) \quad (3)$$

3.5.2 *Drying of agricultural produce:*

The energy requirement Q_v of solar drying plants is a function of the following factors:

- type of material to be dried,
- moisture content of the material to be dried before and after drying,
- maximum permissible drying temperature,
- temperature and relative humidity of the ambient air, and
- drying time.

Therefore the necessary collector area is calculated as follows:

$$A = Q_v \, / \, \vartheta \, . \; Q_i . \; cos \; \alpha \; (m^2)$$

3.6 Wind energy

The bandwidth of the power density of wind is very large. The value of the ten-minute maximum is many times higher than the annual average. This broad variation in the power density of wind causes specific challenges for the choice and construction of wind power generators. The annual value of wind energy is calculated as follows:

$$e_a = \sigma_L \, / \, 2 \int_{t_1}^{t_2} V^3 \, dt \; (J/m^2)$$

e_a	$=$	specific annual energy (J/m^2)
σ_L	$=$	air density (kg/m^3)
V	$=$	speed of wind (m/s)
t_1	$=$	beginning of year
t_2	$=$	end of the year

The total efficiency of a wind generator consists of three components:

ϑ	$=$	Cp ϑm ϑAM
C_p	$=$	performance coefficient
ϑ_m	$=$	mechanical efficiency
ϑ_{AM}	$=$	efficiency of linked up machine

The performance coefficient C_P is a measure for the aerodynamic quality of a windmill. The mechanical efficiency registers the losses in the transmission bearings and the gear. The transformation losses in the linked up machine, e.g. generator or pumps, are considered by efficiency ϑ_{AM}. So the efficiency of a wind energy converter is the ratio of the specific efficiency to the maximum theoretical wind potential. Modern wind generators achieve a total efficiency of 40% or more.

The successful exploitation of wind energy is site specific depending on the wind resources of the area being exploited. The economic viability of wind energy converters depends on the wind conditions that prevail at a particular site. Electricity generation from wind energy requires generally a wind speed higher than 4 meters per second (m/s). For wind pumps, lower wind speeds can be sufficient. However, most wind pumps will not start below a wind speed of 3 m/s and will furl at about 12 to 15 m/s.

Figure 23: *Windfarm established in 1985 in coastal region in western Denmark (top); Lagerwey MW-size windmills on mountain top, Mie prefecture, Japan (left); Mie University, Japan, is testing experimental windmill (left).*

From a technical point of view, we can install the wind converter in favourable wind fields at distances of 5 to 8 rotor diameters. At this distance, the direct mutual influencing of the wind converters is still small; this gives a surface factor - defined as proportion of rotor face to field area - of 0.016 or less.

For the installation of wind converters in addition to the existence of reasonable wind speed on the site there are other limiting factors, which must be considered. For example:

- use limitation and prohibitions in:
  - settlements and traffic areas,
  - forest areas,
  - natural and landscape conservation areas,
  - leisure and recreation areas

- required safety distances and limitations from aesthetic aspects.

These restrictions affect turbines of various sizes differently.

On the other hand, the agricultural land use is not limited by windmill installations if the turbines are not too small and are not placed too densely. The area required for foundations does not hinder the activities of fodder or crop production.

Figure 24: Windmills are being installed to allow normal agricultural utilization: Planting of willow shoots in northern Denmark (left); row of big windmills in grass field in Galicia, Spain (right).

3.7   Biomass

Biomass includes all materials of organic origin (e.g. all natural living or growing materials and their residues). The delimitation compared to fossil energy carriers begins with peat, a secondary product of rotting organic matter. Therefore, all plants and animals, their residues and wastes as well as materials resulting by their transformation (paper-cellulose), organic wastes from the food industry, as well as organic wastes of households and industrial production qualify as biomass.

Biomass appears in different forms, which is simultaneously produced in organisms. Cellulose is the most frequent organic substance. Cellulose is a polysaccharide, consisting of pure glucose chains, which have been connected by hydrogen in crystal bounds.

Woody plants consist of 20 to 30% hemicelluloses. It is also a polysaccharide but consists not only of pure glucose chains but other sugars as well.

The wood pulp lignin constitutes about 30% of the woody plants. Lignin causes the lignification of vegetable cells by occlusion into the cellulose matrix. Compared with cellulose and lignin the remaining other form of biomass plays a small role.

Starch (1050 million of t/y), sugar (100 million t/y) as well as fat, protein and dyes (130 million t/y) constitute only 1% of the world biomass production.

Table 6:   Different forms of biomass and their worldwide annual growth

Biomass form	Worldwide annual growth	
	%	billion t/y
Cellulose	65	100
Hemicellulose	17	27
Lignin	17	27
Starch		1
Sugar	1	0,1
Fat		
Protein		0,13
Dyes		
Sum	**100**	**155**

Biomass represents one of the most important renewable energy resources for the future. It is used diversely as an energy carrier (Figure 25).

In thermo chemical procedures, biomass is transferred to secondary energy carriers by oxidation, by application of heat or by chemical processing. In the agricultural sector the biomass can be used as energy resource by combustion, or as fuel and biogas.

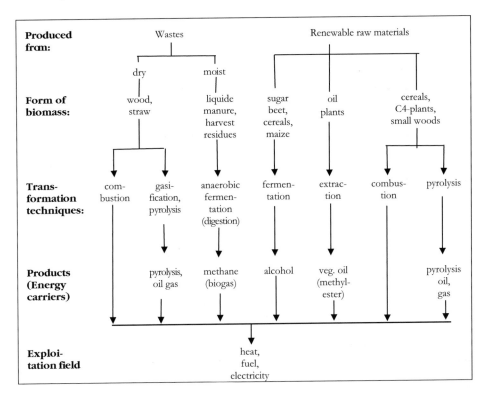

Figure 25: Exploitation of biomass in an energy resource

3.8   Energetic use of Biomass

3.8.1   *Combustion*

The oldest thermo chemical use of biomass is combustion. Almost half of the present forest woodcutting in the world is exploited as an energy resource for heating and cooking. Table 7 shows the part of fire wood consumption of total wood occurrence in industrial and developing countries. In the developing countries people also use other biomass resources e.g. dung or agricultural waste (straw) as combustible materials.

The net energy per unit mass of matter freed during combustion is called the net heating value H_U. For biomass the heating value depends on the specific net heating value of the dry matter (DM), organic dry matter (ODM), and their part (1-x) to the total mass. It also depends on the specific evaporation heat of water.

$$H_U = (1 - x)H_{UTS} - x2{,}441 \text{ MJ/kg (MJ/kg)}$$

H_{UTS} = heating value of biomass dry matter
X   = moist
2,441 MJ/kg   = energy of initial temperature of 25 °C

The net heating value H_U represents the basis for the further calculation of the energetic use of biomass in an Integrated Energy Farm.

The energetic value of biomass as a fuel depends on its humidity content. Rather than using biomass with high humidity content for power generation other procedures are preferable, such as biogas generation.

Figure 26:   Biomass for energy: Truck transporting woodfuel to Bamako, Mali (left); cow dung dried and stacked in India (middle); chipping of forest residues for district heating in Denmark (right).

The table 7 shows that the net heating value of leaves and coniferous woods does not vary much from each other. The heating value depends more on the moisture content than the type of wood.

For absolutely dry wood the combustion temperature reaches approximately 1200°C. It decreases with increasing humidity content, simultaneously increasing the consumption of wood per energy unit produced.

Table 7:   The part of firewood of total wood occurrence

	Wood (Total) million m³	Firewood million m³	Part of Firewood in %
Industrialised	1446	235	16
countries	673	93	14
North America	275	40	14
Western Europe	61	12	20
Eastern Europe	33	3	9
Oceania	355	81	23
Former USSR	48	7.5	17
other			
Developing countries	1983	1594	80.4
Africa	481	437	91
Latin America	409	293	72
Near East	57	42	74
Far East	1027	815	79
other	9	5.8	64

Figure 27:   Efficient use of biomass: A collection cooking stoves from rural Africa (left); mass oven in family house in Germany (middle); wood pellet boiler with automatic feeding for central heating in Styria, Austria (right).

Table 8:   *Dry matter heating value of different type of biomass*
            (Kleemann, M., Meliß, M. 1988)

combustible biomass	heating value H_{uts} der TS MJ/kg
ash (tree)	18,6
beech	18,8
oak	18,3
coniferous woods	19,0
paper wastes	17,0
straw	16,0
sugar cane	15,0
leaves	18,0
fats	39,0
fuel oil	43,0

Wood efficiencies vary significantly depending upon the type of application and combustion stove.

- open light   5 -10%
- simple stove   20 - 30%
- open chimney   10 - 30%
- stove, cooking stove   40 - 50%
- fuel burnout stove (20 - 400 kW)   60 - 70%
- sub-fire stove (20 -1200 kW)   60 - 80%

3.8.2 *Extraction*

Extraction is the second type of physical bioconversion. Direct extraction entails the separation of energy carriers from the biomass e.g. through cold or hot pressing, steam breaking, acid hydrolysis or other procedures. Some plants are able to produce, in addition to the partly oxidized C - H - combinations as cellulose or lignin, oxygen-free hydrocarbons, which can be used directly as vegetable oils or as energy carriers. However, vegetable oils are employed today mainly for food production as well as for the production of enamels, colours, soaps and cosmetics items. The use of vegetable oils for heating and drying as well as for engines is limited by region. An extension of the cultivable areas can lead in some countries to a competitive situation with food production. Above all, as in the developing countries where a food deficit exists, the effort to utilize vegetable oil as fuel can be problematic. Here the principle "food over fuel" should be retained.

Under specific conditions however, in the developing countries, it is regionally possible to set up adapted techniques for extraction of vegetable oil and as well as for their use in engines in order to strive for a largely autonomous, regional economy that is in cycle with nature.

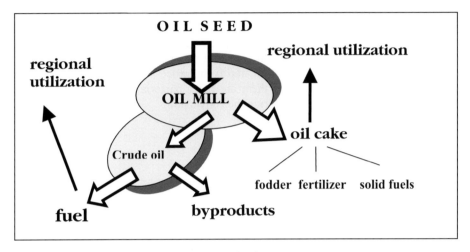

Figure 28: Oil extraction, products, by-products and utilization
(source: Krause 1995)

3.8.3 Biogas production

While the thermal gasification requires biomass having less humidity and the alcoholic fermentation depends on bio raw materials and residues containing sugar and starch, for biogas generation we need liquid or semi liquid raw materials and residues. Here the material composition does not have a great importance. Solid biomass must only have small lignin content and a sufficiently large surface.

Figure 29: Biogas is mature technology, that finds many applications: Farm biogas with digester, cogenera-
tion and gasholder in Thy, Denmark (top); one of several big community biogas plants, also in
Denmark (left); biogas expert from India visits family biogas plant in West Africa (right).

During the biogas generation, different bacteria groups are in interplay with each other and transform organic material under anaerobic conditions. Biogas is a mixture of methane and carbon dioxide.

The formation of methane from organic material occurs in three steps. The two first steps are the preparation steps; the third step is the process of methane production. In each step of the conversion process there are other bacteria influencing the process. These bacteria multiply themselves during the conversion.

The three steps of the process are:

1. Acid formation or hydrolysis,
2. Acid disassembly,
3. Methane production

The quantity of gas produced as well as the methane content essentially depends on following influencing parameters that determine the environment for bacteria:
- type of substrate,
- dry matter,
- temperature,
- retention time,
- pH-value,
- quantity of substrate.

Basic materials for methane extraction can be liquid manure, plant waste, and by-products of food production. (Table 9). All of them are containing decomposable materials such as proteins, fat or carbohydrates (starch or cellulose).

Table 9:   *Biogas production from different biowastes and residues*
Source: Folkecenter for Renewable Energy (2000)

Branch	Waste type	VS [%]	CH$_4$ production per VS [Nm³/kg]	Biogas production per ton biomass [Nm³/ton]
Agriculture	Milking cows	8.5	0.21	27
	Young stock	7.3	0.21	23
	One year sows	6.4	0.29	28
	Porkers	5.8	0.29	24
	Hens	31	0.29	136
	Broilers	45	0.29	195
Pig slaughterhouse	Content of stomach / gut	16-20	0.46	110-138
	Fat and flotation slurry	4.5-35	0.50	34-263
	Remains after sieving	12	0.30	54

Branch	Waste type	VS [%]	CH$_4$ production per VS [Nm^3/kg]	Biogas production per ton biomass [Nm^3/ton]
Cattle slaughter-house	Content of stomach / gut	16	0.40	96
	Fat and flotation slurry	36	0.58	313
	Remains after sieving	12	0.30	54
Poultry slaughter-house	Fat and flotation slurry	7-40	0.61	84-366
Dairies	Flotation slurry	7-8	0.40	42-48
	Whey	4-6	0.33	20-30
Oil mills	Bleaching clay	40	1.00	600
	Misc. organic material	25	0.47	176
Margarine	Fat slurry	90	0.81	1094
Potato flour	Fruit sap	4	0.35	21
Pectin industry	Kelp remains	4-5	0.21	13-16
Brewery	Yeast / dregs	10	0.26	39
	Yeast / dregs	21	0.26	82
	remains from filters	11	0.26	43
Pharmacy industry	Production slurry	5-100	0.30	23-450
Tannery	Glue leather	17	0.50	128
Vegetable market	Sap from vegetables	2.5-5	0.45	17-34
Fish oil / meal industry	Fat and flotation slurry	8-24	0.36	43-136
Fish filleting industry	Fat and flotation slurry	7-20	0.45	47-135
Herring cannery	Miscellaneous slurry	8-11	0.55	66-91
Mackerel cannery	Fat and flotation slurry	17-23	0.55	140-190
Shellfish industry	Fat and flotation slurry	20-26	0.75	225-281
Smoke fish industry	Fat and flotation slurry	8-44	0.59	71-389

The optimal dry matter content of the substrate to be fermented is 5 to 12%. This value can be adjusted by addition of water or urine. An optimal gas production is achieved in a pH- field from 6,5 to 7,2. In general the biogas plants work continuously. A specific quantity of biomass is added daily and a corresponding quantity of fermented substrate is diverted.

Liquid manure e.g. from large scale animal husbandry is an excellent substrate for biogas generation. In contrast, solid manure (with straw mixtures) often leads to considerable technical difficulties. In tables 10, 11 and 12 are presented some data referring to biogas generation:

Table 10:   Composition of animal excrement, referring to dry matter

	pig excrement in %	cattle excrement in %	Chicken excrement in %
carbohydrates	48	20	25
fat	4	4	4
protein	19	15	29
crude fiber	20	40	15
ash	19	21	27

Table 11:   Quantity of animal excreta per day

quantity	cattle per unit	per CU	pig per unit	per CU = 5 units	chicken per unit	per CU = 250 units
liquid manure (kg/d)	50	50	4	20	0,1	25
dry matter (Kg ODM/d)	5	5	0,4	2	0,03	7,5
quantity (l gas/Kg ODM)	300	300	400	400	400	400
quantity (l gas/d)	1500	1500	160	800	12	3000
methane content in %	60	60	70	70	70	70
heating value (Mj/Nm³ Gas)	20	20	23	23	23	23

CU = cattle unit = 500 kg life weight
OTM = Organic Dry Matter

Table 12:   Quantity of gas and retention time of agricultural products

Material	Biogas m³/kg ODM	retention time (d)
wheat straw	0,367	78
sugar beet leaves	0,501	14
potato tops	0,606	53
maize tops	0,514	52
clover	0,445	28
grass	0,557	25

An important factor for a biogas plant working continuously is the space load (**Rb**). It specifies, how much (kg) organic dry matter may be loaded per day in a cubic meter of fermenter volume.

$$R_b = m_{su}\, C_{OTM} / V_R \left[(kg/d)/m^3 \right]$$

m_{su}:   daily supply with substrate (kg/d)
C_{OTM}:   concentration of organic dry matter
V_R: reactor volume (m³)

When the reactor is filled, we have approximately the following situation:

$$V_R = V_{SU} \ (m^3)$$

V_{SU}:   volume of substrate (m³)

Where C_{OTM} is:

$$C_{OTM} = m_{OTM} / m_{su} = V_{OTM}\, \mu_{OTM} / V_{SU}\, \mu_{SU}$$

μ: density (kg/ m³)
m: mass (kg)
V: volume (m³)

The fermentation time of the substrate t_{vw} in days can be calculated as follows:

$$t_{vw} = V_{su} / V_{su} = m_{su} / m_{SU}$$

V_{su}:   Volume of daily supply of substrate (m³ / d)

Decreasing the space load increases the fermentation time. The optimal duration of fermentation e.g. for mesophilic bacteria is between 20 to 30 days. For example, if we have a substrate density of 1000 kg / m³ and a concentration of 0.08, we will have a daily dry matter supply of 2.7 to 3 kg per m³ fermenter volume corresponding to the quantity of 33.3 to 37.5 kg substrate.

The cumulative gas production of animal excrement conducts at 30 °C is 0.380 to 0.400 m³ of biogas per kg ODM. That corresponds to a daily gas gain per m³ fermenter volume of 1.026 to 1.2 m³.

The production conditions are strongly influenced by the type and the composition of the substrate.

4 Planning of Integrated Energy Systems for Rural Communities

4.1 Scenario 1

The necessary data such as climatic conditions, soil data as well as the data concerning the identification of production factors are normally available in an existing agricultural region. This data should be taken as basis for the planning of Integrated Energy Farms and Systems.

However, the energy demand of an agricultural area cannot be covered alone by the use of circuit economy e.g. by energetic use of agricultural wastes and residues or through solar and wind energy. In some climatically unfavourable regions we need, in addition to available resources, the production of energy raw materials (biomass) on the farm. However, this will lead to a contentious situation concerning land use between farm products and energy raw materials. The production of the energy crop must not be introduced or expanded at the expense of the other farm products such as foods.

Figure 30:   Renewable energy for rural development: Premixing of cow manure for biogas, Aligarh, India (left); jatropha plant oil crop in West Africa (middle); biogas cook stove in family kitchen, Mali (right).

The process of planning an integrated energy system at the community level should include optimising the farm production in such a way that reserves (land, capital, manpower) are freed from the field of food production, and can be utilized in an economically feasible way in the process of production of energy and energy raw materials.

A comparison between industrialized and developing countries illustrates that the preferences are very different. In industrialized countries, agriculture is under pressure to reduce food production, while in developing countries food production is of highest importance. Therefore, in the industrialized countries, the replaced production capacity should be used preferably for the production of marketable energy and energy raw materials in order to create new sources of income for the farmers. On the other hand, in the developing countries, the additional production reserves of the farm should only serve to secure energetically autonomous food production.

For planning integrated systems, the following points should be first considered:

- Intensification and optimisation of existing agricultural production considering site conditions. This must increase the yields, productivity and income (**optimisation model**).

- Choice of adapted production branches with consideration of existing demand and elimination of economically inefficient production branches.

- Introduction of sustainable production models with the aim of minimization of input (capital, surface) in case of unchangeable net profits.

- Analysis of additional inputs (capital, land, manpower) with regard to its use for additional food production or for production of energy raw material. (**Preference model**)

- Integration of energy into agricultural production with regard to regional preferences should consider the following aspects:

  o criteria of economical profitability
  o technical and economical feasibility
  o availability of technologies, integration possibilities of different technologies
  o surface, capital and manpower requirements
  o socio-economic and other regional particularities

- Installation of an energy production and user management system

The Figures 31 and 32 show the different steps of planning and implementation:

4.2 Scenario 2

An integrated energy farm could be created on the basis of the energy and food requirement of a specific number of persons. Assuming the farm area is available, the necessary site data such as climate, soil etc. should be first determined. This can be recorded from national or regional statistics, calculated or measured.

In the arid and semi arid regions irrigation possibilities must be determined. Here, an analysis of the relevant irrigation system and its economic feasibility is very important taking into consideration the regional sociological aspects.

The Figures 33 and 34 show a graphical presentation of the planning and implementation steps:

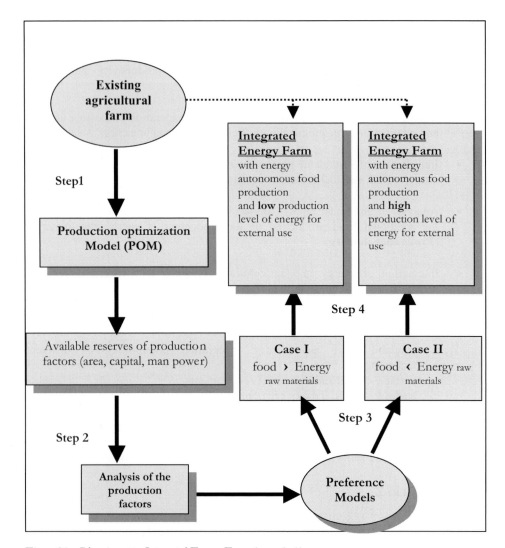

Figure 31: Planning steps Integrated Energy Farm (scenario 1)

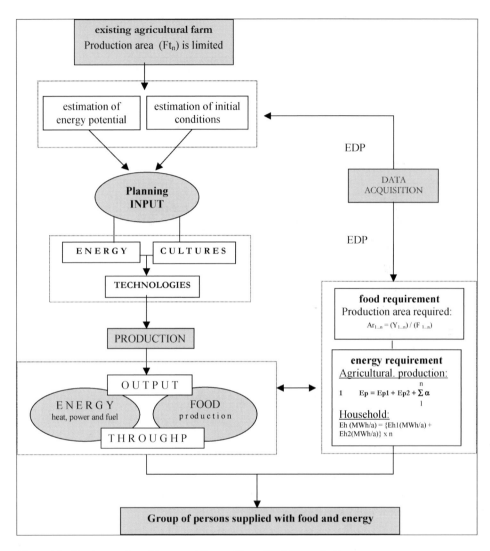

Figure 32:   Implementation of Integrated Energy Farm, IEF (Scenario 1)

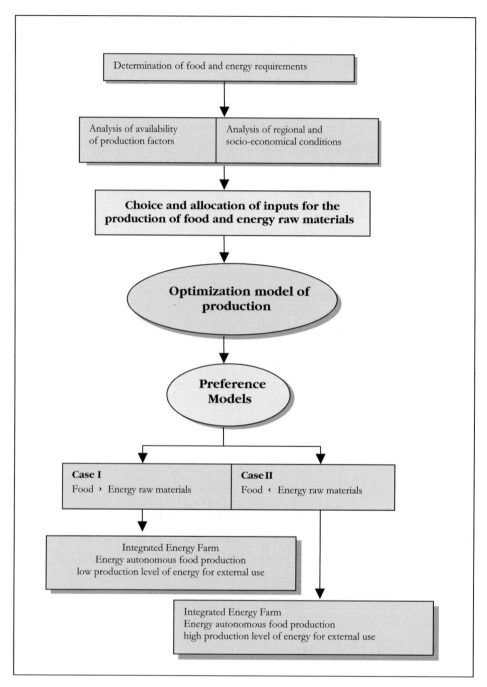

Figure 33: Planning Steps of IEF (scenario 2)

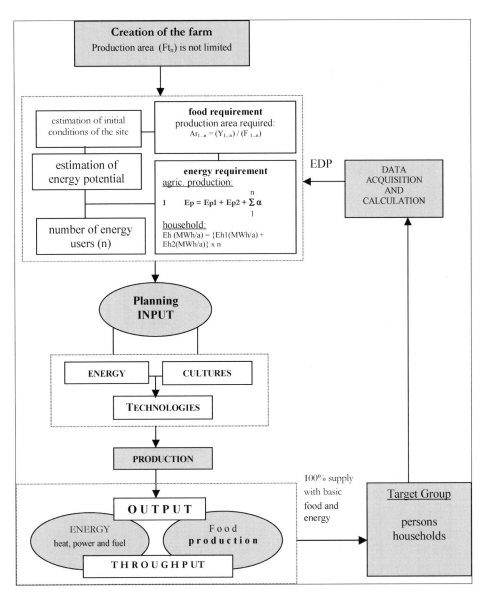

Figure 34:   *Implementation of the Integrated Energy Farm, IEF (scenario 2)*

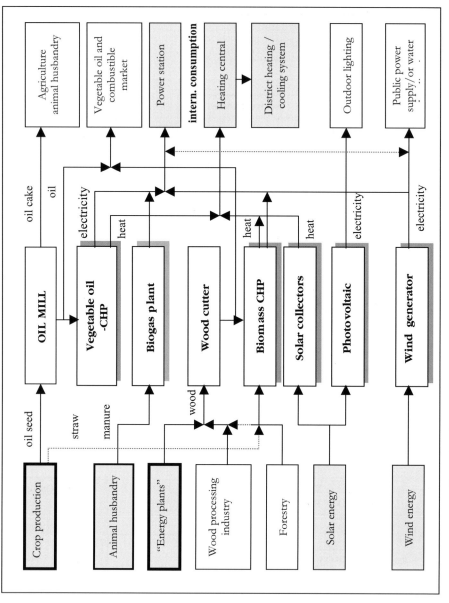

Figure 35: General system complex of Power Generation on an Integrated Energy Farm.

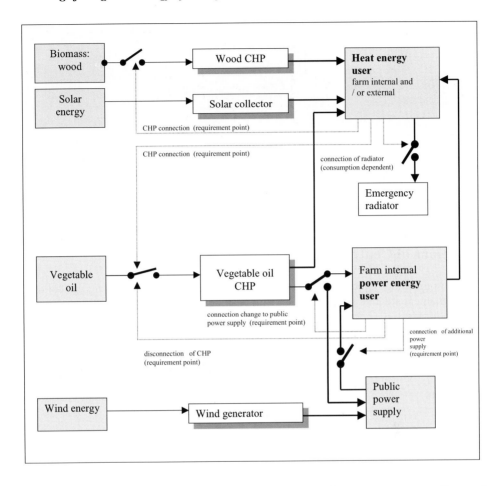

Figure 36: Basic diagram of the system control unit of the Power Generation system on an Integrated Energy Farm. (Source: Wolf, M. (1998) modified by the authors)

Figure 37: Energy-efficient combined heat and power: CHP district heating utility with 2x3000 kw gas motors. for 6000 residents of Faaborg, Denmark (left); 180 kw cogeneration using ELSBETT plant oil technology (middle); CHP for district heating in Hilpoltstein, Germany, fuelled by local rape seed oil (right).

4.3   Case study I: Implementation of IEF under climatic conditions of Central Europe

The main planning objectives of the implementation of IEF on an existing agricultural farm settlement are that it should be autonomous in energy supply.

4.3.1   Specifications:

Climate: Climatic conditions Northern Germany
Farm size: 100 ha
Production: Crops, vegetables, horticulture, energy plantation animal husbandry, vegetable and fruit production

Distribution of the farm area

Crops and root plants:	60 ha
Oil plants:	10 ha
Vegetables:	5 ha
Fruit trees:	5 ha
Grassland	17 ha
Building:	3 ha
Total:	100 ha

The infrastructure of the farm could consist of a residential building with administration tract, office rooms (area: 400 m^2), storage rooms, workshops, greenhouses and stable facilities.

4.3.2　Farm production

The activities of the created integrated energy farm consist of:

<u>Crop production:</u>　　　　　　　　70 ha
<u>Energy plantation:</u>　　　　　　　　　　15 ha
<u>Grassland /fodder:</u>　　　　　　　2 ha

<u>Horticulture</u>
- Vegetables:　　　　　　　5 ha
- Fruit trees:　　　　　　　5 ha

<u>Animal husbandry</u>
- Sheep breeding (No.)　　　100
- Chicken (No.)　　　　　　　　500

The following table presents the production data of different agricultural activities:

Table 13:　Agricultural production per year (estimated)

Production branch	Products	Area	Quantity
Crop production	Cereals (wheat and barley)	40.0 ha	100 t
	Sunflower	10.0 ha	15 t
	Potatoes	20.0 ha	3000t
Vegetables	Different products	5.0ha	100 t
Fruit trees	Apples, pears, etc.	5.0 ha	200 t
Animal husbandry	Hens (eggs)	-	75.000 units
	Sheep breeding*	2.0 ha	2.0 t LW

* 1.0 Lamb per year, LW = Life Weight, dt = 0.1 ton

4.3.3　Energy requirement

a)　Administration and household

The area for the living and administration buildings is projected to be approx. 400 m^2. The determination of the **heat energy requirement** is based on the following criteria:

- The farm buildings should be well insulated so that relatively low energy for heating will be required corresponding to 60-70 Watt per m^2 reference area.
- The total hours needed for heating will be appx. 1800 h/y (10 hours/d x 6 months /y x 30 days).

> **Heat requirement (kWh/y) = used area (m^2) x 70W/ m^2 x 1,800 h**
> **= 400 x70 x1800 x 10^6 = 50.4 MWh/y**

Hot water requirement is calculated to be 40 l/d/person. The annual demand for energy for hot water from 20 to 60 °C is calculated as follows:

$$40 \times (60 - 10) \times 4.19 \times 365 \times 0.000278 = 850.3 \text{ kWh/y/person}$$

For the entire building, heating load of approx. **28.0 kW** should be available, so that the entire annual energy demand will be 50.4 **MWh/year**.

The following table illustrates the data of electricity needed for various activities in administration and household (6 persons):

Table 14: Power requirement in MWh/year

Fields	Connection value (KW)	Full load hours (h)	Consumption (MWh/y)
Basic load	2.5	8,760	21.9
Household	2.5	1,100	2.75
Cooking	8.2	730	5.99
Administration	1.5	1,300	1.95
Workshop	12.1	450	5.45
Sum	**26.8**		**38.04**

The basic load of 2.5 kW for the entire farm is projected for permanent (24 h /day) power generation for circulating pumps, ventilation, engines, refrigerator, emergency lighting etc.

The total electricity consumption amounts to 38.1 *MWh/year*, this corresponds a consumption of *104.2 KWh* per day.

b) Agricultural activities

Energy requirement, subdivided in power, heat and fuel for the various farming activities is summarized in the following table.

For all farming activities 824 tractor hours per year are needed with a total fuel consumption of 6476 litres. The requirement of electricity and heat energy is calculated to 39.5 MWh/a and 75.0 MWh/y, respectively.

Summarizing the energy demand of the farm production and that of the household and administration space, the following energy is required for the entire farm:

Heating energy:	**125.4 MWh/y**
Electricity:	**77.6 MWh/y**
Fuel:	**6476.2 Litres**

Table 15: Electricity, heat energy and fuel requirement on the farm

Branches	Activities	Area in ha	Th*/ area	Electricity (MWh)	Heat (MWh)	Fuel (litres)
Cereals	soil preparation	40.0	144.0	-	-	1166.4
	care		72.0	-	-	583.2
	harvest		10.0	-	-	81.0
	straw transport /recovery		64.0	-	-	518.4
Potatoes	soil preparation	19.0	95.0	-	-	76.5
	care		95.0	-	-	769.5
	harvest/storage		91.2	5.5	-	738.7
Oil plants	soil preparation	10.0	67.0	-	-	542.7
	care		16.0	-	-	129.6
	harvest/storage		20.0	9.8	-	162.0
	processing		-	7.5	-	-
Vegetables	care	5.0	6.5	-	-	52.7
	harvest		10.0	-	-	81.1
	processing		-	7.4	-	-
Fruit trees	care	5,0	7.5	-	-	60.8
	harvest/transport		6.0	-	-	48.6
	storage		-	3.5	-	-
Sum 1	-	79.0	704.2	33.7	-	5704.2
Animal husbandry	feeding	5.0	120.0	-	-	972.0
	other	-	-	5.8	75,0	-
Sum 2		5.0	120.0	5.8	-	972.0
Total		**84.0**	**824.2**	**39.5**	**75.0**	**6476.2**

* Th= tractor hour

4.3.4   *Energy production on the farm*

The total energy demand calculated for heat was 125.4 MWh/y and for electricity 77.6 MWh/y. The required energy should be produced fully on the farm. Surplus electric energy can be sold to the public energy network.

The energy farm exploits exclusively renewable energy resources. The concept includes a combined use of solar and wind energy as well the energy from biomass. The energy supply system of the farm consists of the followings technologies:

- **Photovoltaic plant** of approx. 5.3 kW (50 modules with an output capacity of 105 Wp/module on a surface of 50 m^2)
- **Thermal solar collectors** with 100 m^2 collecting surface
- **Windmill** of 300 kW,
- **Wood/Biomass-CHP**, 100 kW thermal capacity
- **Stirling engine** with 10 kW electrical and 20 kW of thermal capacity that is operated with biogas
- **Biogas plant** having the production capacity of 750 - 1000 m^3 biogas/year

The following table shows the contribution of various technologies to produce energy from renewable sources:

Table 16:   Estimated annual energy production on the farm

	Heat energy	Electricity
Photovoltaic plant	-	5.0 MWh 1)
Thermal solar collector	40.0 MWh	-
Windmill	-	120.0 MWh 2)
CHP (cogeneration)	260.0 MWh	125.0 MWh 3)
Stirling engine (with biogas)	4.4 MWh	1.6 MWh
T otal	304.4 MWh	251.6 MWh

1) Calculated on the basis of an average global sun radiation of 2.7 KWh/m^2/d and its energetic use of 10-12%.
2) Wind speed on the site = 4.5 – 5.0 m/sec 3) annual operation hours = 2100 h

4.3.5　Origin of biomass

The biomass for energy generation is provided on the farm. 15 ha will be cultivated with fast growing trees species e.g. eucalyptus and poplars, as well as various energy plants such as giant reed, miscanthus, different tall grasses etc.

Consequently, the farm utilizes about 90 tons of biomass annually as energy raw material producing approx. 400 MWh/y.

The biogas plant is oper with animal manure, plant wastes and farm residues (estimated quantity about 40 tons/a) producing 700 - 750 m^3 biogas.

A Stirling engine operated with biogas produces 4.4 MWh/y heat and 1.6 MWh/y of electricity.

> **The total quantity of energy generation from the produced biomass on farm**
> **amounts to 304.4 MWh/y heat energy**

4.3.6　Contribution of different renewable energy sources

The following table shows the contribution of different energy sources in percentage to cover the total energy requirement of the farm (heat: 377.4 MWh/y, power: 107.6 MWh/y, fuel: 7093.4 litres)

Table 17:　Energy demand and energy generation

	Solid biomass	Oil plants	Biogas	Solar-energy	Wind-energy	Total	% of total demand
Heat	400.0 MWh	-	4.4 MWh	40.0 MWh	-	444.4	117.8
Electricity	-	-	1.6 MWh	7.5 MWh	200.0 MWh	209.1	194.4
Fuel	-	2,500 l	-	-	-	2,500 l	35.2

4.3.7　Investment requirement

The investment requirement has been estimated on the basis of information collected from different professional organisations, energy agencies and energy producers. The total sum of the required investment, including all capital and additional costs such as financing and services costs amounts to about 918,000 EUR.

In this case, the estimated total investment requirement consists of the following costs of equipment, installation and service:

Windmill (300kW):	300,000
Photovoltaic solar cells for power generation (50 units with the capacity of 105Wp/units) +installation:	38,000
Solar collectors for heat energy (100 m²):	15,000
Biomass - CHP:	125,000
Biogas plant with reservoir and generator:	75,000
Oil mill + tank + Installation:	110,000
Wood cutter + Container:	35,000
Plantation of energy plants:	125,000
Financing costs, planning and service etc.:	95,000

Total	EUR:	918,000

4.4   Case study II: Arid and semi arid regions

The main objectives of the planning are the implementation of a farm settlement that is autonomous in energy supply including desalination and cooling facilities from renewable energy sources.

4.4.1   Specifications:

Climate: Arid and Semi Arid Regions and Islands
Farm size: 100 ha
Production: crops, vegetables, horticulture, energy plantation, animal husbandry, pisciculture and apiculture

Distribution of the farm area	
Crops and root plants:	60 ha
Oil plants:	10 ha
Vegetables:	5 ha
Fruit trees:	5 ha
Grassland:	17 ha
Building:	3 ha
Total:	**100 ha**

The infrastructure of the farm could consist of a residential building with administration space, office rooms (using area: 400 m²), storage rooms, workshops, greenhouses and stable facilities.

4.4.2 Farm production

The activities of the created integrated energy farm consist of:

Crop production:	70 ha
Energy plantation:	15 ha
Grassland /fodder:	2 ha
Horticulture	
• Vegetables:	5 ha
• Fruit trees:	5 ha
Animal husbandry	
• Sheep breeding (No.)	100
• Chicken (No.)	500

The following table presents the production data of different branches:

Table 18: *Agricultural production per year (estimated)*

Production branch	products	area	quantity
Crop production	Cereals (Wheat and barley)	40.0 ha	100 t
	Sunflower	10.0 ha	15 t
	Potatoes	20.0 ha	3000 t
Vegetables	Different products	5.0ha	100 t
Fruit trees	Apples, Pears, etc.	5.0 ha	200 t
Animal husbandry	Hens (eggs)	-	75,000 units
	Sheep breeding*	2.0 ha	2.0 t LW

* 1.0 Lamb per year, LW = Life Weight , dt = 0.1 ton

4.4.3 Energy requirement

a. Administration and household

The area for the living and administration buildings is projected to be appr. 400 m^2.
The determination of **cooling load requirement** is based on the following criteria:

- The farm buildings should be well insulated so that relatively low energy for cooling will be required which corresponds to 1 ton per 20 m^2 (3.5 kW/ton).
- The total hours needed for cooling will be appr. 4320 h/a (18 hours/d x 8 months /y x 30 days).

> Formula:
> **Cooling load requirement (MWh/y) = area (m²) x 3.5 kW/ 20 x 4320 h**
> **= 400 x3.5/20 x4320 = 302.4 MWh/y of heat energy to be used in absorption**
> **chillers to produce chilled water for cooling purposes**

Hot water requirement is calculated to be 40 l/d/ Person. The annual demand for energy for hot water from 20 to 60 °C is calculated as follows:

> 40 x (60 – 20) x 4.19 x 365x 0.000278 =783.0 kWh/y/person

For all the buildings, a cooling load of approx. **75.0 kW** should be available, so that the entire annual energy demand will be **302.4 MWh/year**.

The following table indicates the data of electricity needed for various activities in administration and household (6 persons):

Table 19:   Power requirement in MWh/year

fields	connection value (KW)	full load hours (h)	consumption (MWh/y)
basic load	2.5	8,760	21.90
household	2.5	1,100	2.75
cooking	8.2	730	5.99
administration	1.5	1,300	1.95
workshop	12.1	450	5.45
sum	**26.8**	!Syntax Error,)	**38.04**

The basic load of 2.5 kW for the entire farm is projected for permanent (24 h /day) power generation for circulating pumps, ventilation, engines, emergency lighting, refrigerators, etc.
The total electricity consumption amounts to 38.1 *MWh/year*; this corresponds to a consumption of *104.2 KWh* per day.

b. Agricultural activities

Energy requirement, subdivided in power, heat and fuel for the various farming activities is summarized in the following table.

Table 20:   Electricity, heat energy and fuel requirement on the farm

branches	activities	area in ha	Th*/ area	electricity (MWh)	heat (MWh)	fuel (liter)
cereals	soil preparation	40.0	144.0	-	-	1166.4
	care		72.0	-	-	583.2
	harvesting		10.0	-	-	81.0
	straw transport/ recovery		64.0	-	-	518.4
potatoes	soil preparation	19.0	95.0	-	-	769.5
	care		95.0	-	-	769.5
	harvest/storage		91.2	5.5	-	738.7
oil plants	soil preparation	10.0	67.0	-	-	542.7
	care		16.0	-	-	129.6
	harvest/storage		20.0	9.8	-	162.0
	processing		-	7.5	-	-
vegetables	care	5.0	6.5	-	-	52.7
	harvest		10.0	-	-	81.1
	processing		-	7.4	-	-
fruit trees	care	5.0	7.5	-	-	60.8
	harvest/transport		6.0	-	-	48.6
	storage		-	3.5	-	-
energy plantation		15.0				
grassland		2.0				
sum 1	-	79.0	704.2	33.7	-	5704.2
animal	feeding	5.0	120.0	-	-	972.0
husbandry	other	-	-	5.8	75.0	-
sum 2		5.0	120.0	5.8	-	972.0
total		84.0	824.2	39.5	75.0	6476.2

* Th= tractor hour

For all farming activities 24.2 tractor hours per year is needed with a total fuel consumption of 6476.2 liter. The requirement of electricity and heat energy is calculated to 39.5 MWh/a and 75.0 MWh/y, respectively.

Summarizing the energy demand of the farm production and that of the household and administration space, so we will have the following energy requirement situation for the entire farm:

heating energy:	**377.4 MWh/y**
electricity:	**107.6 MWh/y**
fuel:	**7093.4 Litres**

4.4.4   *Energy production on the farm*

The total energy demand was calculated for heat at 377.4 MWh/y and for electricity 107.6 MWh/y. The required energy should be produced fully on the farm. The surplus electric energy will be used to power the desalination system.

The energy farm should exploit exclusively renewable energy resources. The concept includes a combined use of solar and wind energy as well as the energy from biomass.

The energy supply system of the farm consists of the followings technologies:

- **Photovoltaic plant** of approx. 5.3 kW (50 modules with an output capacity of 105 Wp/module on a surface of 50 m^2)
- **Thermal solar collectors** with 100 m^2 collecting surface
- **Windmill** of 300 kW,
- **Wood/Biomass-CHP**, 100 kW thermal capacity
- **Stirling engine** with 10 kW electrical and 20 kW of thermal capacity that is operated with biogas
- **Biogas plant** having the production capacity of 750 - 1000 m^3 biogas/year

The following table shows the contribution of various technologies to produce energy from renewable sources:

Table 21:   Estimated annual energy production on the farm

	heat energy	electricity
Photovoltaic plant	-	7.5 MWh 1)
Thermal solar collector	40.0 MWh	-
Windmill	-	200.0 MWh 2)
CHP(cogeneration)	360.0 MWh	
Stirling engine (with biogas)	4.4 MWh	1.6 MWh
Total	**404.4 MWh**	**208.1 MWh**

1) calculated on the basis of an average global sun radiation of 5,0 KWh/m^2/d and its efficiency of 10-12%.
2) wind speed on the site = 3.5 – 4.0. m/sec
3) annual operation hours for CHP = 3100 h

4.4.5   Origin of biomass

The biomass for energy generation is provided on the farm. 15 ha will be cultivated with fast growing tree species e.g. eucalyptus and poplars as well as various energy plants such as giant reed, miscanthus, different tall grasses etc.

Consequently, the farm disposes of about 90 tons of biomass annually as energy raw material producing approx. 400 MWh/y.

The biogas plant is operates with stable manure, plant wastes and farm residues (estimated quantity about 40 tons/a) producing 700 - 750 m^3 biogas.

A Stirling engine operated with biogas produces 4.4 MWh/y heat and 1.6 MWh/y of electricity.

> **The total quantity of energy generation from the produced biomass on farm amounts to 404,4 MWh/y heat energy**

4.4.6   Contribution of different renewable energy sources

The following table shows the contribution of different energy sources in percentage to cover the total energy requirement of the farm (heat: 377,4 MWh/y, Power: 107,6 MWh/y, fuel: 7093,4 Liter)

Table 22:   Energy demand and energy generation

	Solid biomass	Oil plants	Biogas	Solar-energy	Wind-energy	Total	% of total demand
heat	400.0 MWh	-	4.4 MWh	40.0 MWh	-	444.4	117.8
electricity	-	-	1.6 MWh	7.5 MWh	200.0 MWh	209.1	194.4
fuel	-	2,500 l	-	-	-	2,500 l	35.2

4.4.7 Investment requirement

The investment requirement has been estimated on the basis of information collected from different professional organisations, institutions of energy supply and producers. The total sum of the required investment, including all capital and additional costs such as financing and services, amounts to about **1,118,000 EUR**.

In this case, the estimated total investment requirement consists of the following costs of plants, installation and service:

1.	Windmill(300kW):	300,000
2.	Photovoltaic solar cells for power generation (50 units with the capacity of 105Wp/units) +Installation:	38,000
3.	Solar collectors for heat energy (100 m^2):	15,000
4.	Biomass - CHP:	125,000
5.	Biogas plant with gas storage and generator:	75,000
6.	Oil press + tank + Installation:	110,000
7.	Wood cutter + container:	35,000
8.	Plantation of energy plants:	125,000
9.	Financing costs, planning and service etc. :	9,.000
10.	Reverse Osmosis desalination unit (capacity 10 m3/h)	100,000
11.	Absorption chillers and other air conditioning equipment	50,000
12.	Irrigation equipment	50,000

Total :	**EUR**	**1,118,000**

5 Renewable Energy Resources and Technologies

5.1 Biomass and Bioenergy

Biomass is the term used to describe all the organic matter that exists on the earth's surface produced by photosynthesis. The source of all energy in biomass is the sun with the biomass acting as a kind of chemical energy store. Biomass is constantly undergoing a complex series of physical and chemical transformations--and being regenerated while giving off energy in the form of heat to the atmosphere. To make use of biomass for our own energy needs we can simply tap into this energy source. In its simplest form a basic open fire is used to provide heat for cooking, warming water or warming the homes.

More sophisticated technologies exist for extracting this energy and converting it into useful heat or power in an efficient way. The exploitation of energy from biomass has played a key role in the evolution of mankind. Until relatively recently it was the only form of energy which was usefully exploited by humans and it is still the main source of energy for more than half the world's population for domestic energy needs.

5.1.1 Herbaceous Energy Crops
Herbaceous energy crops are perennials that are harvested annually after taking two to three years to reach full productivity. These include such grasses as seitchgrass, miscanthus (also known as Elephant grass or e-grass), bamboo, sweet sorghum, tall fescue, kochia, wheatgrass, and others.

Figure 38: Plantation with fast growing pines in Michigan, USA (left); biomass grown on grey waste water for the removal of mineral nutrients (right).

5.1.2 Woody Energy Crops
Short-rotation woody crops are fast growing hardwood trees harvested within five to eight years after planting. These include hybrid poplar, hybrid willow, silver maple, eastern cottonwood, green ash, black walnut, sweetgum, and sycamore.

5.1.3   Industrial Crops

Industrial crops are being developed and grown to produce specific industrial chemicals or materials. Examples include kenaf and straws for fiber, and castor for ricinoleic acid. New transgenic crops are being developed that produce the desired chemicals as part of the plant composition, requiring only extraction and purification of the product.

5.1.4   Agricultural Crops

These feedstocks include the currently available commodity products such as cornstarch and corn oil, soybean oil and meal, wheat starch, other vegetable oils, and any newly developed component of future commodity crops. They generally yield sugars, oils, and extractives, although they can also be used to produce plastics and other chemicals and products.

Figure 39:   Sorghum bicolor (left)
Figure 40:   Miscanthus giganteus (right)

Figure 41:   Miscanthus giganteus *and genotypes*
Figure 42:   Willow field in the FAL (El Bassam 2000)

Figure 43: Bamboo-forest in China

Figure 44: Arundo donax in Greece

Figure 45: Biodiversity is an economical necessity for cultivated forests

Figure 46:   Oilfields of the 21st century (El Bassam 2002)

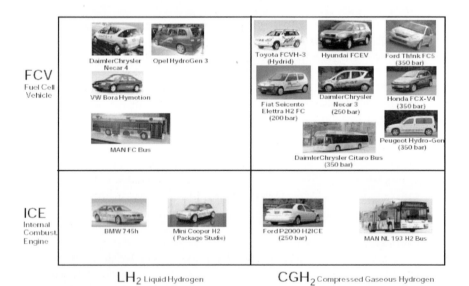

Figure 47:   Hydrogen activities, worldwide

5.1.5   Aquatic Crops

A wide variety of aquatic biomass resources exist such as algae, giant kelp, other seaweed, and marine microflora. Commercial examples include giant kelp extracts for thickeners and food additives, algal dyes, and novel biocatalysts for use in bioprocessing under extreme environments.

Figure 48: Microalgae growth chambers (El Bassam 2000)

5.1.6   Agricultural Crop Residues

Agriculture crop residues include biomass, primarily stalks and leaves, not harvested or removed from the fields in commercial use. Examples include corn stover (stalks, leaves, husks and cobs), wheat straw, and rice straw. With approximately 80 million acres of corn planted annually, corn stover is expected to become a major biomass resource for bioenergy applications.

5.1.7   Forestry Residues

Forestry residues include biomass not harvested or removed from logging sites in commercial hardwood and softwood stands as well as material resulting from forest management operations such as pre-commercial thinnings and removal of dead and dying trees.

5.1.8   Municipal Waste

Residential, commercial, and institutional post-consumer wastes contain a significant proportion of plant derived organic material that constitute a renewable energy resource. Waste paper, cardboard, wood waste and yard wastes are examples of biomass resources in municipal wastes.

5.1.9   Biomass Processing Residues

All processing of biomass yields byproducts and waste streams collectively called residues, which have significant energy potential. Residues are simple to use because they have already been collected. For example, processing of wood for products or pulp produces sawdust and collection of bark, branches and leaves/needles.

Figure 49:   Organic residues from industry and agriculture collected for incineration at district heating plant (left); mountain of wood chips is awaiting shipment from Chile to Japan (right).

5.1.10   Animal Wastes
Farms and animal processing operations create animal wastes that constitute a complex source of organic materials with environmental consequences. These wastes can be used to make many products, including energy.

5.1.11   Landfill Gas
Landfill gas consists mainly of methane and CO_2, is produced when organic wastes decay in landfill sites. The methane content means that it has potential as fuel, either to generate electricity or to provide process heat.

5.2   Regional Availability
Natural biomass resources vary in type and content, depending on geographical location. For convenience sake, we can split the world's biomass producing areas into three distinct geographical regions:

5.2.1   Temperate regions
Temperate regions produce wood, crop residues such as straw and vegetable leaves, and human and animal wastes. In Europe short rotation coppicing (SRC) has become popular as a means for supplying woodfuel for energy production on a sustainable basis. Fast growing wood species, such as willow are cut every two to three years and the wood chipped to provide a boiler fuel. There are also many non-woody crops which can be grown for production of biofuels and biogas, and investigation of energy crops for direct combustion is underway (Table 23). In western countries, where large quantities of municipal waste are generated, this is often processed to provide useful energy either from incineration or through recovery of methane gas from landfill sites.

Table 23:   *Energy plant species (Temperate climate)*

• Cordgrass (*Spartina spp.)*	• Reed Canary Grass (Phalaris arundinacea.*)*
• Fibre sorghum *(Sorghum bicolor)*	• Rosin weed *(Silphium perfoliatum)*
• Giant knotweed (*Polygonum sachalinensis)*	• Safflower (*Carthamus tinctorius)*
• Hemp (*Cannabis sativa)*	• Soy bean (*Glycine max)*
• Kenaf (*Hibiscus cannabinus)*	• Sugar beet (*Beta vulgaris)*
• Linseed (*Linum usitatissimum)*	• Sunflower (*Helianthus annuus)*
• Miscanthus *(Miscanthus x giganteus)*	• Switchgrass *(Panicum virgatum)*
• Poplar *(Populus spp.)*	• Topinambur *(Helianthus tuberosus)*
• Rape *(Brassica napus)*	• Willow *(Salix spp.)*

5.2.2   Arid and semi-arid regions

Arid and semi arid regions - produce very little excess vegetation for fuel. People living in these areas are often the most affected by desertification and often have difficulty finding sufficient woodfuel. Some of energy crops which could be grown in these regions are listed in table 24.

5.2.3   Humid tropical regions

Humid Tropical regions - produce abundant wood supplies, crop residues, animal and human waste, commercial, industrial and agro- and food-processing residues. Rice husks, cotton husks and groundnut shells are all widely used to provide process heat for power generation, particularly. Sugarcane bagasse is processed to provide ethanol as well as being burned directly; and many plants, such as sunflower and oil-palm are processed to provide oil for combustion. Many of the world's poorer countries are found in these regions which are rich in energy plant species (Table 25) and hence there is a high incidence of domestic biomass use. Tropical areas are currently the most seriously affected by deforestation, logging and land clearance for agriculture.

Table 24:   Energy plant species (Arid and Semiarid climate)

• Argan tree (*Argania spinosa)*	• Olive (Olea europaea.*)*
• Broom (Ginestra) *(Spartium junceum)*	• Poplar *(Populus spp.)*
• Cardoon (*Cynara cardunculus)*	• Rape (*Brassica napus)*
• Date palm (*Phoenix dactylifera)*	• Safflower (*Carthamus tinctorius)*
• Eucalyptus (*Eucalyptus spp.)*	• Salicornia (*Salicornia bigelovii)*
• Giant reed (*Arundo donax)*	• Sesbania (*Sesbania spp.)*
• Groundnut (*Arachis hypogaea)*	• Soybean *(Glycine max)*
• Jojoba *(Simmondsia chinensis)*	• Sweet sorghum *(Sorghum bicolor)*

Table 25:   Energy plant species (humid climate)

• Aleman Grass (*Echinochloa polystachya)*	• Jatropha (*Jatropha curcas.)*
• Babassu palm *(Orbignya oleifera)*	• Jute *(Crocorus spp.)*
• Bamboo (*Bambusa spp.)*	• Leucaena (*Leucaena leucoceohala)*
• Banana (*Musa x paradisiaca)*	• Neem tree (*Azadirachta indica)*
• Black locust (*Robinia pseudoacacia)*	• Oil palm (*Elaeis guineensis)*
• Brown beetle gras (*Leptochloa fusca)*	• Papaya (*Carica papaya.)*
• Cassava *(Manihot esculenta)*	• Rubber tree *(Acacia senegal)*
• Castor oil plant *(Ricinus communis)*	• Sisal *(Agave sisalana)*
• Coconut palm (*Cocos nucifera)*	• Sorghum *(Sorghum bicolor)*
• Eucalyptus *(Eucalyptus spp.)*	• Soybean *(Glycine max)*
	• Sugar cane (*Saccharum officinarum)*

5.2.4   Current and Potential Uses of Biomass

Woodfuels consist of three main commodities: fuelwood, charcoal and black liquor. Fuelwood and charcoal are traditional forest products derived from the forest, trees outside forests, wood-processing industries and recycled wooden products from society. Black liquors are by-products of the pulp and paper industry. Table 26 shows regional production of fuelwood (including wood for direct use as fuel and for conversion into charcoal) in 1999, about 1.4 billion tonnes of fuelwood were produced worldwide, which is about 470 mtoe or about 5% of the world total energy requirement.

FAO is making strenuous efforts to improve the quality and quantity of wood energy data, in particular by ensuring that as far as possible production and consumption statistics cover all sources of fuelwood - not only established forests, but also other wooded land, farms and gardens, roadside trees, etc. Black liquor supplied about 72 mtoe of energy in 1997. Thus, it can be roughly estimated that woodfuels in total contribute about 540 mtoe annually to the world energy requirement.

On average, the annual per-capita consumption of woodfuels is estimated to be 0.3-0.4 m³ or around 0.1 toe, but with considerable regional variances.

Table 26:   Total fuelwood production in 1999

	mtoe	%
Africa	141.1	29.9
North America	38.5	8.1
South America	37.7	8.0
Asia	216.1	45.8
Europe	34.9	7.4
Middle East	0.2	0.0
Oceania	3.8	0.8
TOTAL WORLD	**472.3**	**100.0**

The amount of woodfuel use varies considerably among regions, mainly owing to differences in stages of development. Fuelwood use is especially common in the rural areas of developing countries as the main source of household energy, while charcoal is mainly used by urban and peri-urban dwellers. In general terms, fuelwood production can be assumed to be more or less equal to fuelwood consumption within a region. However, the same rule cannot be applied to the amount of fuelwood used for charcoal making. In fact, the production of 1 tonne of charcoal requires approximately 6 m³ of wood.

Asia is by far the largest producer and consumer of fuelwood, accounting for 46% of world production. Africa has the second highest share at 30%, followed by South America and North America, both at about 8%. On the other hand, the production and consumption of black liquor are concentrated in developed countries with large pulp and paper industries. Therefore, about 50% of black liquor consumption is in North America, followed by Europe with 19% and Asia with 12%. Africa is the most intensive user of woodfuels in per-capita terms, with an average annual per-capita consumption of 0.77 m³, or 0.18 toe. In Africa, almost all countries rely on wood to meet basic energy needs. The share of woodfuels in African primary energy consumption is estimated at 60% to 86%, with the exception of North African countries and South Africa. On average, about 40% of the total energy requirement in Africa is met by fuelwood.

In Asia, about 7% of the total energy requirement is met by fuelwood and the per-capita consumption level is not very high; however, the situation varies from country to country. Many countries in South and South East Asia, such as Nepal, Cambodia, Thailand and Indonesia, rely heavily on fuelwood, consuming more than 0.5 m3 per capita annually. In Latin America, about 10% of the total energy requirement is met by fuelwood.

In Europe and North America, the share of fuelwood in the total energy requirement is low, at 1.2% and 1.4% respectively. However, for countries such as Finland, Sweden, the USA and Canada, per-capita consumption is quite high if black liquor is included. In Austria, Finland and Sweden, wood energy provides about 12% to 18% of the country's total primary energy supply.

5.2.5   Households
Woodfuels, as well as other traditional sources of energy such as agricultural residues and animal dung, have an important role in the lives of the rural populations in developing countries. Fuelwood and charcoal, the commonest forms of woodfuel, are used widely as household power sources in poor rural neighbourhoods in developing countries. In Pakistan and the Philippines, for example, fuelwood supplied 58% and 82% respectively, of rural household energy consumption in the late 1980's to early 1990's. The major energy end-use is cooking in households: about 86% of fuelwood consumed in urban households in India is for this purpose, while the rest is used mainly for water heating.

Figure 50:   Utilisation of biomass in West Africa: Mini-truck with woodfuel load is heading for Bamako, Mali (left); woman gets bag of oil-rich jatropha nuts weighed at the village plant-oil pressing station (right).

In Africa, in 1994, more than 86% of total woodfuel consumption was attributed to the household sector. Dependence on woodfuels to meet household energy needs is especially high in most of sub-Saharan Africa, where 90% to 98% of residential energy consumption is met by woodfuels. In the European Union, most woodfuels are used by households, which account for around 60% of the total wood energy consumed.

5.2.6   *Industry*

Most of the non-household fuel wood consumption occurs in agro-based rural industries such as crop drying, tea processing and tobacco curing, as well as in the brick and ceramic industries. Woodfuel consumption by such users is smaller than that of households; nevertheless, it should not be overlooked as it can constitute 10% to 20% of fuel wood use in some Asian countries. In Africa, in 1994, it was estimated that traditional industries accounted for about 9.5% of woodfuel consumption.

Woodfuels are also used in larger-scale industries, mostly in the form of charcoal. For example, in Brazil some 6 million tonnes of charcoal are produced every year for use in heavy industry, such as steel and alloy production.

The widespread use of fuel wood and charcoal is attributable to various reasons. Fuel wood is often the cheapest and most accessible form of energy supply in the rural areas of developing countries. In many cases, it is harvested at no monetary cost as a common property resource from forests and from scattered pockets or belts of trees along field margins or roadsides and on waste or common ground. Both fuel wood and charcoal are traded, mainly in and around urban areas or beside transportation routes. Charcoal is the favoured commodity for trading, as it burns more efficiently than fuel wood and is easier to transport and store.

Rising income levels and expanding urbanisation usually make it possible for people to have access to more modern forms of energy. The rapid expansion of conventional energy capacity to meet the energy needs of industries and modern lifestyles results in a reduced share of woodfuels in the total energy mix, as well as lower per-capita consumption of woodfuels.

In developed countries, biofuels (including woodfuels) are mostly used for electricity and heat generation in cogeneration systems (combined heat and power production) on industrial sites or in municipal district heating facilities. In Oceania, North America and Europe, black liquors are widely used for fuelling the heat and power plants of the large pulp and paper industries. Almost all of their energy needs are met by black liquors and, in some cases, surplus electricity is sold to the public grid.

Recent trends in both energy and environmental policies, mainly in developed countries, promote the use of woodfuels. In many countries, deregulation, liberalisation and privatisation of energy markets over the past two decades have stimulated competition among energy suppliers and have presented new opportunities for non-fossil energy sources. Technological developments in woodfuel production, transportation, combustion, etc. are helping to make woodfuels more cost-competitive. In addition, woodfuels are increasingly receiving more attention for the environmental benefits they provide. Some countries have raised the taxes on fossil fuels, thus encouraging a decrease in the use of these fuels and, in some cases, increased use of other energy sources. Moreover, several countries and regions, for instance, Canada, Finland, Denmark, Austria, Germany and the European Community, have adopted energy policies aimed at an increased use of woodfuels.

5.2.7   *Woodfuels*
Woodfuels come from a variety of supply sources, such as forests, non-forest lands and forest industry by-products. In 1998, 3.2 billion m3 of wood were harvested worldwide, more than 50% of which was used for woodfuel. It has often been said that most wood-fuels are obtained from forests, contributing to deforestation in a major way. However, it is now estimated that considerable amounts of woodfuels come from non-forest areas, such as village lands, agricultural land, agricultural crop plantations (rubber, coconut, etc.), homesteads and trees along roadsides. In some Asian countries, the proportion of woodfuels originating from non-forested areas exceeds 50%.

In some areas, nevertheless, woodfuel consumption exceeds the sustainable production from available and accessible supply sources. In Haiti, the Andean highlands and the Sahelian countries, as well as around large cities such as Khartoum, Bamako and Dakar, the obvious pressure on forest resources is causing concern.

5.2.8   *Peat*
Peat is a soft organic material consisting of partly decayed plant matter together with deposited minerals. For land to be designated as peat land, the depth of the peat layer, excluding the thickness of the plant layer, must be at least 20 cm on drained, and 30 cm on undrained land.

The energy content of in-situ peat depends on its moisture and ash contents. However, the organic component of peat deposits has a fairly constant anhydrous, ash-free calorific value of 20-22 MJ/kg, and if the total quantity of organic material is known, together with the average moisture and ash contents, then the peat reserve may be equated with standard energy units.

During the past decade the Greenhouse Gas (GHG) problem has become a major issue in discussions concerning the environmental impacts of energy production. In this debate the peat industry has been the loser, because peat is classified as a fossil fuel and CO_2 emissions released during its combustion are taken into account in full in the calculations of the International Panel for Climate Change (IPCC).

5.2.9   *Production, Handling and Logistic*

Production Improvements
Improvements in agricultural practices will lead to increased biomass yields, reductions in cultivation costs, and improved environmental quality. Key elements include new plant genetics and breeding technology, new analytical techniques and evaluation techniques, and the development of tools to enable precision agriculture, such as remote sensing and geographic information systems (GIS).

Material Handling
Materials handling systems for biomass constitute a significant portion of the capital investment and operating costs of a bioenergy conversion facility. Requirements depend on the type of biomass to be processed as well as the feedstock preparation requirements

of the conversion technology. Biomass storage, handling, conveying, size reduction, cleaning, drying, and feeding equipment and systems are included.

Collection, Logistics and Infrastructure

Harvesting biomass crops, collecting biomass residues, and storing and transporting biomass resources are critical elements in the biomass resource supply chain.

5.2.10 Future Bioenergy Scenarios

Projections of future energy scenarios present various possibilities in terms of the magnitude of woodfuel use in the future. The World Energy Outlook 2000, prepared by the International Energy Agency, projects an increase in the consumption of combustible renewables and waste (CRW; including fuel wood, charcoal, crop residues and animal wastes) between 1997 and 2020 in absolute terms in every region of the world.

In developing countries, the primary energy supply through CRW will grow from 886 mtoe in 1997 to 1.103 mtoe in 2020, at an annual growth rate of 1%. However, the share of CRW in the total primary energy supply in developing countries will drop from about 24% in 1997 to 15% in 2020, owing to a more rapid expansion of commercial energy use as a result of rising income levels. In Africa the share would still remain high, at around 43% in 2020, based on a projection of relatively modest increases in income levels within the region.

In the recently published Intergovernmental Panel on Climate Change (IPCC) Special Report on Emissions Scenarios, it is estimated that the largest renewable energy potentials in the medium term (to 2025) shall be found in the development of modern biomass (70 to 140 EJ), followed by solar (16 to 22 EJ) and wind energy (7 to 10 EJ). In the longer term, the maximum technical energy supply potential of biofuels is estimated to be 1,300 EJ, second to solar with a potential of around 2,600 EJ. However, the report points out some constraining factors, such as competition with agriculture for food production, productivity in biomass production, etc.

Biomass other than wood includes agricultural and wood/forestry residues and herbaceous crops grown specifically for energy but excludes forest plantations grown specifically for energy. Currently there are a number of dedicated energy plantations, e.g. Brazil, where there are about 3 million ha of eucalyptus plantations used for charcoal making; China, which has a plantation programme for 13.5 million ha of fuel wood by 2010; Sweden, where there are about 16,000 ha of willow plantations used for the generation of heat and power; and the USA, where some 50,000 ha of agricultural land has been converted to woody plantations, possibly rising to as much as 4 million ha (10 million acres) by 2020. But all current plantations have tended to follow traditional agricultural and forestry practices.

Municipal solid waste (MSW) is potentially a major source of energy. However, there are a number of reasons why this source of biomass will not be considered in this commentary, which comprises many different organic and non-organic materials.

Difficulties and high costs associated with sorting such material make it an unlikely candidate for renewable energy except for disposal purposes. Re-used MSW is mostly for recycling, e.g. paper. MSW disposal would be done in landfills or incineration plants.

It is well known that biomass is a very poorly documented energy source. Indeed lack of data has hampered sound decision-making when it comes to biomass energy. For example, a close examination of this chapter's Country Notes illustrates the variations and discrepancies between the biomass resources reported by the WEC Member Committees quite well. The inability to fully address the indigenous biomass resource capability and its likely contribution to energy development is still a serious constraint to the full realisation of this energy potential, despite a number of efforts to improve biomass energy statistics.

Previous commentaries estimated, roughly, that biomass consumption in rural areas of developing countries (including all types of biomass and end-uses) was about 1 tonne (15% moisture, 15GJ/t) per person/year and about 0.5 tonne in semi-urban and urban areas. This assumption is still generally valid today. It seems that while in relative terms traditional biomass energy consumption may be declining in some parts of the world, in absolute terms the total amount of biomass energy is increasing. There are many variations due to the large numbers of factors involved, such as availability of supply, climatic differences, population growth, socio-economic development, cultural factors, etc.

The increasing interest in biomass for energy since the early 1990's is well illustrated by the large number of energy scenarios showing biomass as a potential major source of energy in the 21st century. Hoogwijk et al (2001) have analysed 17 such scenarios, classified into two categories: i) Research Focus (RF) and ii) Demand Driven (DD). The estimated potential of the RF varies from 67 EJ to 450 EJ for the period 2025-2050, and that of the DD from 28 EJ to 220 EJ during the same period. The share of biomass in the total final energy demand lies between 7% and 27%. For comparison, current use of biomass energy is about 55 EJ.

Biomass resources are potentially the world's largest and most sustainable energy source - a renewable resource comprising 220 billion oven-dry tonnes (about 4 500 EJ) of annual primary production (Hall & Rao, 1999). The annual bio-energy potential is about 2900 EJ, though only 270 EJ could be considered available on a sustainable basis and at competitive prices. The problem is not availability but the sustainable management and delivery of energy to those who need it.

Residues are currently the main sources of bio-energy and this will continue to be the case in the short to medium term, with dedicated energy forestry/crops playing an increasing role in the longer term. The expected increase of biomass energy, particularly in its modern forms, could have a significant impact not only in the energy sector, but also in the drive to modernise agriculture, and on rural development.

The most reasonable approach would be to concentrate efforts on the most promising residues from the sugar cane, pulp and paper, and sawmill industrial sectors. More than 300 million tonnes of bagasse are produced worldwide, mostly used as fuel in sugar cane factories. FAO data show that about 1 248 million tonnes of cane was produced in 1997. About 25% is bagasse, representing some 312 million tonnes. The energy content of one tonne of bagasse (50% moisture content) is 2.85 GJ/tonne cane milled. This excludes barbojo (top and leaves) and trash - representing the largest energy fraction of the sugar cane (55%) - which is currently mostly burned off or left to rot in the fields. This large potential is thus currently almost entirely wasted.

Forestry residues obtained from sound forest management can enhance and increase the future productivity of forests. One of the difficulties is to estimate, with some degree of accuracy, the potential of residues that can be available for energy use on a national or regional basis, without more data on total standing biomass, mean annual increment, plantation density, thinning and pruning practices, current utilisation of residues, etc. Recoverable residues from forests have been estimated to have an energy potential of about 35 EJ/y (Woods & Hall, 1994).

Figure 51: AFPRO training and demonstration facility for family biogas in Aligarh, U.P., India

A considerable advantage of these residues is that large parts is generated by the pulp, paper and sawmill industries and thus could be readily available. Currently, a high proportion of such residues is used to generate energy in these industries, but there is no doubt that the potential is considerably greater. For example, Brazil's pulp and paper industry generates almost 5 mtoe of residues that is currently largely wasted. The estimated global generation capacity of forestry residues is about 10 000 MWe.

The potential of energy from dung alone has been estimated at about 20 EJ worldwide (Woods & Hall, 1994). However, the variations are so large that figures are often meaningless. These variations can be attributed to a lack of a common methodology, which is the consequence of variations in livestock type, location, feeding conditions, etc. In addition, it is questionable whether animal manure should be used as an energy source on a large scale, except in specific circumstances like biogas fermentation that will not influence negatively on the fertilizer value of the residue.

Energy crops can be produced in two main ways: as dedicated energy crops in land specifically devoted to this end and intercropping with non-energy crops. Energy forestry/crops have considerable potential for improvement through the adoption of improved management practices. It is difficult to predict at this stage what the future role of biomass specifically grown for energy purposes will be. This is, in many ways, a new concept for the farmer, which will have to be fully accepted if large-scale energy crops are to form an integral part of farming practices.

In the past decade a large number of studies have tried to estimate the global potential for energy from future energy forestry/crop plantations. These range from about 100 million ha to over a billion ha, e.g. Hall et al (1993) estimated that as much as 267 EJ/y could be produced from biomass plantations alone, requiring about one billion hectares. However, it is highly unlikely that such forestry/crops would be used on such a large scale, owing to a combination of factors, such as land availability, possible fuel versus food conflict, potential climatic factors, higher investment cost of degraded land, land rights, etc. The most likely scenario would be at the lower end of the scale, e.g. 100-300 million ha.

5.2.11   Gender and Health

Fuel wood collection for household consumption, usually a task for women and children, is becoming more burdensome as fuel wood becomes scarcer. It is estimated that the proportion of rural women affected by fuel wood scarcity is 60% in Africa, nearly 80% in Asia and nearly 40% in Latin America. Moreover, gathering fuel wood can consume one to five hours per day for these women.

The direct burning of fuel wood in poor-quality cooking stoves can result in incomplete combustion, emitting pollutants such as carbon monoxide, methane and particles in the kitchen. In most cases, women are responsible for cooking, spending many hours in the kitchen and thus being more exposed to these pollutants than men are. In addition, the daily hauling of the fuel wood collected imposes a huge physical strain on women.

5.2.12   Upgrading of Biomass

Charcoal

There are several methods for processing wood residues to make them cleaner and easier to use as well as easier to transport. Production of charcoal is the most common. It is worth mentioning at this point that the conversion of woodfuel to charcoal does not increase the energy content of the fuel. Charcoal is often produced in rural areas and transported for use in urban areas. The wood is heated in the absence of sufficient oxygen and full combustion does not occur. This allows pyrolysis to take place, driving off the volatile gases and leaving the carbon or charcoal. The removal of the moisture means that the charcoal has much higher specific energy content than wood. Other biomass residues such as millet stems or corncobs can also be converted to charcoal.

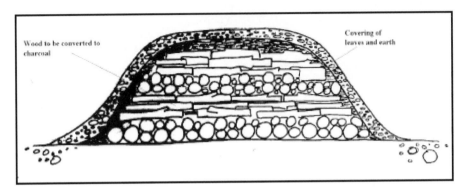

Figure 52:   Traditional Earth Kiln for Charcoal Production

Charcoal is produced in a kiln or pit. A typical traditional earth kiln (see Figure 52) will comprise the fuel to be carbonised, which is stacked in a pile and covered with a layer of leaves and earth. Once the combustion process is underway the kiln is sealed. The charcoal can be removed when the process is complete and cooling has taken place.

Figure 53:   Charcoal Kiln, Kenya ©Heinz Muller/ ITDG

A simple improvement to the traditional kiln is also shown in Figure53. Chimney and air ducts have been introduced which allow for a sophisticated gas and heat circulation system-and with very little capital investment a significant increase in yield is achieved.

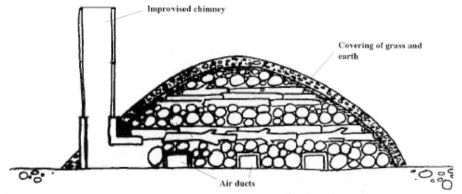

Figure 54: Improved Charcoal Kiln Found in Brazil, Sudan and Malawi

Briquettes

Briquetting is the compression of loose biomass material. Many waste products, such as wood residues and sawdust from the timber industry, municipal waste, bagasse from sugar cane processing, or charcoal dust are briquetted to increase compactness and transportability. Briquetting is often a large-scale commercial activity and often the raw material will be carbonised during the process to produce a usable gas and also more user-friendly briquette.

Pellets

Pellets are more highly compressed loose biomass material than briquettes (ca. 650 kg/m³, diameter 6 mm, length 30-40 mm, ash contents 1% and water contents less than 10%). Pellets are an important and rapidly growing biofuel for the production of heat and electricity. Pellets are developing quickly in the coming years, to the benefit of the environment and local economies. Increased utilisation of wood, agricultural residues, straw and industrial by-products in the Nordic countries, all over Europe and in North America indicate pellets have proven to be a realistic alternative to fossil fuels.

The substitution of fossil fuels with pellets, for heating of buildings and co-generation of electricity, can play an important role in fulfilling the Kyoto commitments for a secure supply of energy.

The potential for expansion of the pellet industry is significant. The Swedish Pellet Producers Association has estimated a market in Europe of 4-5 million tonnes per year within the next 5 years. Pellets will make an important contribution to the world strategy for renewable energy sources.

Figure 55: Small animal draught briquetting press for rice husk, cotton shells and similar for Sudan village (left, middle); cooking stove using briquettes (right); various shapes of pellets and briquettes (below);.

5.2.13 Extraction and Conversion Technologies

Landfill gas

The technology for harnessing landfill gas is well established. The gas is collected from the site through gas wells, consisting mostly of perforated plastic pipes drilled into the waste.

The wells are joined together by plastic pipe work, which is connected to a suction pump to extract the gas. The gas is cleaned and burnt off in a flare stack or combusted as a fuel for cogeneration or other energy purposes. Landfill sites vary widely in the amount of gas that they produce, as the size of the site, the moisture content of the waste and other factors can affect production. Over the last 10-20 years, the technology for electricity generation from landfill gas has evolved into modular units that can be installed as complete packages. Project costs have decreased as installed capacity has increased over the last 15 years.

Figure 56: Landfill site adjacent to the Salvesen Brickworks, Cheshire, UK landfill gas is used to heat the kilns. (Courtesy of ETSU)

Biogas
Anaerobic digestion (AD) involves the breakdown of organic waste by bacteria in an oxygen-free environment. It is commonly used as a waste treatment process but also produces a methane-rich biogas with 50 to 70% being methane and the rest CO_2 and small quantities of other trace elements. Biogas can be used to generate heat and /or electricity.

Figure 57: Main components of large community biogas plant: Digesters (left); gas holder (right); gas engine for CHP by Zantingh (bottom).

AD equipment consists, in simple terms, of a heated digester tank, a gasholder to store the biogas, and a gas-burning engine/generator set, if electricity is to be produced. The organic waste is fermented in the digester at temperatures between 25 and 55 degrees C. The fermentation process is in itself unlike composting to produce heat. The rate of breakdown depends on the nature of the waste and the operating temperature.

The biogas has a calorific value typically between 50% and 70% that of natural gas and can be combusted directly in modified natural gas boilers or used to run internal combustion engines. Apart from biogas, the process also produces a residue that may be separated into liquid and solid components. The liquid fraction can be used as a fertiliser and the solid element may be used as a soil conditioner or further processed to produce a higher value organic compost.

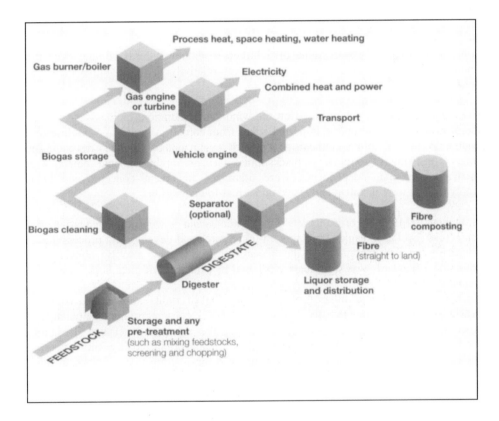

Figure 58: Overview of the anaerobic digestion process.

5.3 Deployment of biogas: Denmark as a case study

Biogas was first introduced in Denmark in the late 1970's. The first plants were by present standards rather primitive and built at farms by dedicated and enthusiastic individuals searching for alternatives to the fossil oil that was being imported from the Middle East. As a consequence of the precarious situation in 1974, the energy policy of Denmark became for almost three decades a concerted effort uniting normally divided parties in broad consensus-with the primary purpose to reduce the dependency on oil.

In 1977 oil covered 95% of the total energy consumption of Denmark. 25 years later oil represents only 40% of the energy consumption of 20 million TOE/year. In the meantime there came significant improvements in energy efficiency combined with increased use of renewable energy sources and extensive application of cogeneration-decentralized solutions being the main elements in the energy policy. The result is an energy consumption at the same level as in 1980.

The positive long-term consequences of the policy in 2002 is a sector employing 30,000 people and an annual export value of 4 billion € with two thirds coming from the windmill industry. Hardly any other country, seen relativily achieved similar industrial benefits by restructuring the energy sector.

Renewable energy equipment based on the wind, solar radiation, biomass and similar natural resources, could not be imported either from abroad in the form of well tested, documented and reliable equipment. At the local level teams of voluntary development groups consisting of engineers, blacksmiths, farmers, architects and others with practical skills and independency were organized to invent concepts and build prototypes that were installed and tested under real-life conditions.

Throughout the 1970's a broad variety of experiments and concepts emerged at the non-commercial level as a result of this bottom-up research and development process that is historically unique. Undoubtedly the success and efficiency also by conventional professional standards was significant, furthered by principles of local cooperation and openness.

Absence of patents and other forms of protection of rights allowed exchange of knowledge between inventors, coordinators, potential consumers and manufacturers. Experiences and knowledge was exchanged at the informal level but also more systematically transferred in manuals and seminars organized by the emerging renewable energy organizations satisfying the need for best practice solutions. Around 1980 the first regular commercial initiatives of manufacturing and marketing of renewable energy equipment within wind power, solar thermal panels, wood stoves and biogas installations were demonstrated. Apart from energy planners and a few physicists, professionals from the universities in general did not see the development of the renewable energies as an earnest and promising effort.

Figure 59: Biogas digester construction details: Slowly rotating agitator (left); 200 mm mineral wool insulation and steel weather protection (below); manure outlet pipe and drive train for agitator (right).

Origins of three farm biogas concepts

Biogas was initially only for individual farms. Its specific development grew out of the process described above involving a broad variety of concepts and experiments. Three parallel development lines contributed especially to advance the technology:

A. One came as a result of a state programme setting up three extremely different digester concepts none of which, however, became of commercial relevance. The government financed the programme, called STUB. The aim was to combine theoretical skills, prototype building, capacity building at the farms, deployment of biogas in agriculture and setting the standards for future biogas plants. Within the project there was published a number of reports mostly based on experiences and measurements from the prototypes.

B. The second biogas development team, Højbogaard, was the combined efforts of a progressive farmer and the local blacksmith. They quickly obtained high gas yields from well-designed digesters in different versions made from steel-plate and concrete by systematically investigating the biological process.

Under the trade name of Bigadan, in 1982 they became suppliers of a scaled-up version for some of the first community biogas plants in Denmark.

C. The third significant biogas development initiative grew out an effort of mobilizing the local human and technological potential in the region of Thy in North Western Denmark. Funding from UNESCO made it possible very early in the biogas implementation phase in 1977 to invite a few international biogas experts with proven knowledge and experiences within a new technology. It became an early and concrete example of international technology transfer within renewable energy, however, in the form of South-North cooperation.

This opportunity allowed consultancy and technology transfer from experts like Ram Bux Singh from India, former leader of the Gobar Gas Research Institute. He had the authority to set important standards in the early development process in terms of valuable technology inputs, by encouraging the farmers in the region and by stressing that comprehensive application of biogas was a long-term learning process and not merely a matter of hard-ware. The Folkecenter has carried on the research for Renewable Energy since 1983. During Ram Bux Singh's stay the first prototypes were installed by local farmers involving the creativity and coordination by the existing renewable energy development team composed of engineers, craftsmen, farmers, teachers and economists.

Materials and technology in biogas construction
The suppliers of biogas plants in Germany and other countries are primarily using concrete digesters, whereas almost all biogas plants in Denmark manufactured are from steel plate. Concrete was also used in Denmark in the 1980's for digester tanks made from concrete prefabricated sections originally developed for liquid manure storage tank.

Figure 60: Farm biogas plant. Belongs to Tom and Jens Kirk, Tousgaard, Denmark.
Digester capacity is 200 m³; the Caterpillar 90 kWₐ gas engine delivers annually 600.000 kWh of electricity into the grid. The heat from the engine is used for heating stables and the family farmhouses.

These tanks had a large diameter in comparison to the height (ratio 3:1) providing a wide span cover structure. Gas leaks, however, occurred and it was determined that gas molecules were diffused through the porous concrete. Therefore steel became the preferred construction material for digesters irrespective of its higher costs. Steel tanks either were welded constructions or enamel steel plate elements bolted together with flexible sealing between the joints. Similar types of tanks are widely used for grain silos.

In Denmark you find three main categories of biogas plants:

- community biogas plants, each 540-7500 m³, delivering electricity to the grid and heat to the town from 50-70% manure and 30-50% industrial waste.
- large, rather primitive farm biogas plants using a concrete manure storage tank as digester, with cogeneration and no co-fermentation.
- Smedemester (Black-Smith) farm biogas with steel digesters, 37-45°C operating temperature, 150-1000 m³ digesters, 2-5% fish waste oil co-fermentation, Cogeneration units of 100-500 KW.

Figure 61: Lemvig: Large community biogas plant
The monthly biogas production is 400-500,000 m³, which by pipeline is delivered to the district-heating cooperative of Lemvig. In a Bergen Ulstein gas engine the biogas is converted into approx.1,000 MWh of electricity and 1,200 MWh of heat every month.

The first community biogas plants were built in the early 1980's. This was part of a desire to supply villages and townships with biogas for cogeneration in the regions of the country that were not going to be supplied by the new Danish fossil gas distribution network. It was anticipated that daily operation of the community biogas plants would become the responsibility of well-trained staff with insight into the biogas process-that individual farmers having a wide variety of duties to take of could not match.

After 1985 more stringent environmental legislation concerning storage and land application of animal manure was introduced. Nine months of storage capacity became a legal requirement for Danish farms. Quantities of nitrogen and phosphates applied per hectare of land also became subject to public regulation. This increased farmer's interest in centralized plants with manure storage capacity included in the concept. The centralized biogas plants all co-digest animal manure and organic waste primarily from the food industry. In 1991 a biogas production level of 30-35 cubic meter of biogas per cubic meter of slurry was evaluated as being a necessary precondition for the economic viability of centralized biogas plant. This production could only be achieved by co-digesting animal manure with wastes from the food industry, normally by admixing 10-30% organic wastes depending on the character, biogas potential, procurement costs and availability of the waste.

The animal manure for community biogas plants is transported from a number of farms to the plants by tank truck or tractor-trailer. After digestion, the slurry is returned to the farm as a nutritionally defined fertilizer, partly to the farms that delivered the fresh manure and partly sold as organic manure to farms engaged in crop farming only.

The volume of animal manure and industrial wastes daily delivered into the community biogas digesters range from 50 to 500 tonnes with resulting biogas production of 1 000 up to 15 000 cubic meters per day.

At farm biogas plants the gas is normally combusted at the farm more or less inde-pendent of the energy requirements of the specific farm. In that respect siting of community biogas plants ensure that the biogas production can be utilized with high efficiency in cogeneration systems with a total efficiency up to 85% when the heat is used for district heating of towns, villages and rural communities.

When installing farm biogas plants, export of heat for district heating is generally not possible. Therefore high overall efficiencies can only be obtained in the event case the biogas farm has the necessary requirements for utilisation of the generated heat.

In the future availability of surplus heat generation will certainly lead to Integrated Energy Farms (IEF), using biogas and other forms of renewable energy technologies, energy storage systems and settlements for the occupants of the complex-in innovative forms of zero emission, ecological villages depending on renewable resources of local origin.

Community biogas plants in their present form include environmental and agricultural benefits, investment savings for the farmers, improved fertilization efficiency, reduced green-house gas emission and cheap, environmentally sound waste recycling of organic material that is generally delivered from the agriculture or fishery to the food industry.

By returning the waste to the fields from which it originally came it helps to maintain the humus balance of the top soil as well as substituting chemical fertilizers produced from fossil fuels. The main disadvantage of centralized plants compared to farm biogas plants is the significant cost of manure transportation.

Admixing is predominantly considered a major advantage for both biogas plants and waste suppliers but has, however, also given rise to uncertainties. Often the long-term reliability of industrial waste supplies has been unpredictable.

There are important aspects related to spreading of diseases with community biogas plants. There is a risk of pathogens from one farm being transferred to other farms by the trucks visiting several farms every day-and due to mixing of manure from many places of origin pathogens from one farm can be spread to other farms.

The veterinarian authorities are of course aware of such risks and have implemented regulations to prevent the problems, however, at considerable costs: Every time a tank truck departs for collecting a new load of manure it has to be carefully cleaned.

In order to prevent pathogens passing through the community biogas plant without being killed, all the slurry passing through the plant is subject to heat treatment (70°C) for a minimum one hour in order to kill pathogens. Since farm biogas plants do not have this kind of additional expenses, farm biogas plants will have significant investment and operational cost advantages.

Technological development of farm biogas plant
With regard to costs, efficiency, and reliability, new standards have been set with the Smedemester (Blacksmith) biogas technology. Two versions were developed, both with steel tanks: 1) Horizontal tanks manufactured in industrial workshops of sizes from 50 m³ to 300 m³. 2) Vertical tanks built on-site in 500 and 1000 m³ sizes. For agitation, slowly rotating blades were preferred due to low energy consumption. The Folkecenter for Renewable Energy developed the technology and supported manufacturers in Denmark, Germany, Lithuania and Japan. After 1988 various European biogas experts were involved in the development process, thus transferring the experiences gained within the various national farm biogas sectors. Erwin Köberle especially provided essential inputs for the success of the Smedemester technology. Also Dr. Anton Perwanger, Dr. Arthur Wellinger and Ekkehard Schneider made valuable contributions.

Figure 62: Biogas in Vycia Farm Kaunas, Lithuania: Demonstration plant, consisting of 3x300 m³ digesters and 75 kW_el plus 110 kW_el John Deere dual fuel engines

In terms of quantities, community biogas plants in general get their waste supply from non-agricultural suppliers. However, this is mostly waste with low content fermentable material coming from the food industry (slaughter houses, dairies, fish processing), pharmaceutical industry, kitchens in hospitals, hotels, etc. Farm biogas economy benefits significantly from the use of small quantities of high-grade waste oil from the food industry, consisting of fish oil and animal fats and other types of organic industrial waste. The waste oil used contains up to 700-1,000 m³ of biogas per m³ of waste substrate.

Figure 63: Danish biogas in Japan: 200 m³ digester developed by Folkecenter for Renewable Energy, Denmark.

5.3.1   CASE: Plant Oil Crops, Biogas and the Dike-pond System with comprehensive Integration

Full integration of crops for vegetable oil, fodder production and biogas can be obtained in combination with the dike-pond system that is in itself an energy neutral aquaculture system. The system combines production of domestic animals, plants and fish in a balanced ecosystem where the wastes from one level of organic production are the precondition of the following step. The water leaving the system has a low level of organic and mineral nutrients. It can be used as recycled water. Any stable biological system is based upon conservation of energy and matter. This means that the production is the surplus created by the system-being the difference between the products leaving the system and the external supplies.

All natural ecosystems tend to obtain an ecological balance where all organisms find a niche. A viable productive ecosystem adapted to certain nutrients and certain external conditions must follow the same basic principles including conservation of energy and matter. A successful system must be based upon well-adapted organisms in balanced niches and stable external conditions.

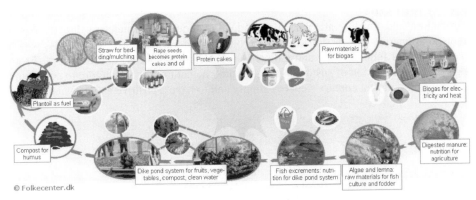

Figure labels in diagram: Straw for bedding/mulching · Rape seeds becomes protein cakes and oil · Protein cakes · Raw materials for biogas · Biogas for electricity and heat · Plantoil as fuel · Compost for humus · Dike pond system for fruits, vegetables, compost, clean water · Fish excrements: nutrition for dike pond system · Algae and lemna: raw materials for fish culture and fodder · Digested manure: nutrition for agriculture · © Folkecenter.dk

Figure 64: Model for a Renewable Energy Farm: Plant Oil Crops, Biogas and Dike-pond System with full Integration

The dike-pond component of the integrated farm system has been derived from the experiences from the delta of the Pearl River in Southern China. Here is an ancient system of cultivation consisting of ponds and dikes. It has high productivity, so it can provide a surplus of a wide range of products to the cities including fish, vegetables, fruits, pigs, eggs and poultry. The production is obtained without any artificial external inputs in the form of chemical fertilizers and pesticides.

With carp ponds as the central element this system, which integrates agriculture and freshwater aquaculture, this is considered one of the most productive cultivating system in the world. The dike-pond system requires very little energy input. Since the waste originating from one production element is perpetually used as input resource for the next element, the system is virtually non-polluting. The system is a complex chain of cultivation elements that is modelled on the cycles of nature.

For this reason, the system is itself a model for the development of local complex cultivation structures in the industrial countries where monocultures and environmental problems make it obvious to seek solutions that respect the natural cycles-and at the same time have high productivity.

It is expected that the manure from one pig can yield 40 kg of carp meat per year and at the same time produce 20 - 30% of the fodder for the pig. The food production per unit area is 4 - 5 times higher than in the traditional agriculture. This intensive biomass production is combined with water purification and energy production.

An advanced farm biogas plant can be connected directly to a greenhouse dike-pond system. This arrangement combines efficient systems of cultivation and energy production. One special advantage is the most efficient use of the nutrients in the manure without any transport. The clean water leaving the system may be used elsewhere on the farm or drained to supplement the groundwater resource.

5.4 Plant Oils

Agriculture and forestry are the only production industries which are directly based upon solar energy. This is done through production of various forms of biomass which can be used for food, directly and indirectly in the form of fodder, and for energy. From a general consideration of sustainability, it would be natural for agriculture to contribute a significant part of a country's energy supply and to have a positive energy balance.

Figure 65:   Production facility for pure plant oil, PPO, at German farm: Oil press and two filters (top); the PPO is ready for use as motor fuel (middle); farm-size oil screw press by Reinhard (bottom).

It would seem obvious to cover the agricultural sectors own energy consumption directly through biomass. As an element in this, the agricultural machinery can be run on plant oil. This would also provide immediate environmental improvements in agriculture.

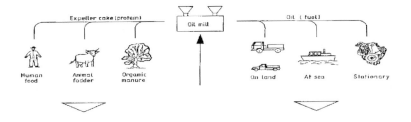

THE ADVANTAGES OF VEGETABLE OIL WHEN USED AS A FUEL

In the countryside	In the country as a whole
Food supply	**Rural mechanisation**
National self-sufficiency in energy	Local and national mechanisation and mobilisation
Increased production	Foreign currency economy
Development of rural infrastructure	Replacement of diesel fossil and other non-renewable fuels
Improved living conditions in the countryside	Less dependence on imported energy
Higher and more secure rural income	Less dependence on electricity supply networks
Prosperity	Industrial development
Creation of jobs in the countryside	Development of rural infrastructure
Increase in the value of the land	Modern technology
Diversity of activities	Ecological advantages
Soil conservation	Increased settlement in countryside

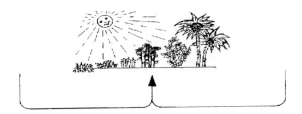

Figure 66: Integration of plant oil in local communities. (Source: ELSBETT)

Plants that produce high amounts of oil are known as oleaginous. Of these only a few are used for commercial purposes. Each climatic region has its own particular oleaginous plants. Tropical regions are especially privileged by a larger variety of plants with higher yields. Most oleaginous plants are significant from an agricultural point of view, and allow for crop rotation and crop combinations. Planting of oleaginous plants makes it possible to combat erosion, reclaim desert land, reforest and manage the soil - allowing a better water control and management.

The main by-product of vegetable oil is expeller cake. Sometimes the cake is the main commercial product and the oil becomes the by-product. The expeller cake is protein rich, can be used to feed humans and animals, and is highly valuable as a natural fertiliser.

Plant oil helps the Integrated Energy Farm (IEF), to become self-sufficient in terms of energy through the use of vegetable oil. In temperate climates the combined cultivation, harvesting and supply of 1000 litres rape seed oil, expeller cake and straw, requires 140-160 litres of oil, that can also be plant oil. The oil is physically extracted by means of a press.

Figure 67:   Applications of pure plant oil, PPO: 30 kW combined heat and power, CHP, with Mercedes Benz motor for farm (left); Toyota truck with converted engine (right); stationary plant oil motor (bottom).

The pressing is cold when the temperature does not exceed 60 °C. The simplicity of the process means that it can be on a large or small scale. The oil has to be cleaned by using a filter, centrifuge or purifier. It is neither explosive nor flammable. It does not emit toxic, carcinogenic gases nor is it harmful to soil, water, animals and humans in the event of accidental spillage.

The solid biofuels, such as straw of the oil plants, can be used for heat production in burners. With more advanced biomass gasifier plants can fuel CHP. Liquid biofuels such as rape seed oil, biodiesel and ethanol are the most versatile fuels because they can easily be used for heat production, CHP, and agriculture mobile purposes that can contribute to a more environmentally sound transport sector - that has shown a lack of positive environmental results and an increasing contribution to energy consumption and greenhouse effects. A breakthrough for plant oil can push this development.

Biodiesel can substitute fossil diesel right away. However, biodiesel presents health and fire hazards in itself, and it is polluting. The pressing and the following esterisation cosrmprises an industrial process with relatively high energy consumption that requires expensives, centralised production plants. In order to meet the requirements of diesel engines, vegetable oils can be modified into vegetable oil ester (transesterification). The transesterification procedure includes the production of methylesters – RME (rape methylester) or SME (sunflower methylester) – and glycerol (glycerine) by the processing of plant oils (triglycerides), alcohol (methanol) and catalysts (aqueous sodium hydroxide or potassium hydroxide). Biodiesel can replace diesel entirely or be mixed with it in different proportions for running diesel engines (which requires modification).

Figure 68: Training courses of diesel engine conversion to PPO with participants from Thailand, Ghana and European countries.

CROP	LATIN NAME	kg/ha/y
Corn	Zea mais	143
Cashew-nut	Anacardium occidentale	148
Oat	Avena sativa	183
Palm	Erythea salvadorensis	189
Lupine	Lupinus albus	195
Rubber seed	Hevea brasiliensis	217
Calendula	Calendula officinalis	256
Cotton	Gossypium hirsutum	273
Soy bean	Glycine max	374
Coffee	coffea arabica	386
Lineseed	Linum usitatissimum	402
Hazel-nut	Corylus avellana	405
Euphorbia	Euphorbia lagascae	440
Pumpkin seed	Cucurbita pepo	449
Coriander	Coriandrum sativum	450
Mustard	Brassica alba	481
Dodder-seed	Camelina sativa	490
Sesame	Sesamum indicum	585
Abyssinian kale	Crambe abyssinica	589
Safflower	Carthamus tinctorius	653
rice	Oriza sativa	696
tung tree	Aleurites spp	790
sunflower	Helianthus annus	801
cocoa	Theobroma cacao	863
peanut	arachis hypogaea	887
Opium poppy	Papaver somniferum	978
rape	Brassica napus	999
olive tree	Olea europaea	1019
Indaia palm	Attalea funifera	1112
gopher plant, spurge	euphorbia lathyris	1119
castor bean	Ricinus communis	1188
bacury	Platonia insignans	1197
jojoba	Simmondsia chinensis	1528
babassu palm	Orbignya martiana	1541
purging nut	Jatropha curcas	1588
macadamia nut	macadamia terniflora	1887
Brazil nut	Bertholletia excelsa	2010
avocado	Persea americana	2217
coconut	Cocos nucifera	2260
oiticica	Licania rigida	2520
Buriti palm	mauritia flexuosa	2743
„Pequi"	Caryocar brasiliense	3142
Macahuba palm	Acrocomia spp	3775
Oil palm	Elaeis guineensis	7061

5.5   Ethanol

Ethanol is ethyl alcohol obtained by sugar fermentation (sugarcane), or by starch hydrolysis or cellulose degradation followed by sugar fermentation, and then by subsequent distillation. The fermentation of sugar derived from agricultural crops using yeast to make alcohol, followed by distillation, is a well-established commercial technology. Alcohol can also be produced efficiently from starch crops (wheat, maize, potato, cassava, etc.). The glucose produced by the hydrolysis of starch can also be fermented to alcohol.

Bioethanol can be used as a fuel in different ways: directly in special or modified engines as in Brazil or in Sweden, blended with gasoline at various percentages, or after transformation into ethyl tertio butyl ether (ETBE). In the first case it does not need to be dehydrated, whereas blends and ETBE production need anhydrous ethanol.

Moreover, ETBE production requires neutral alcohol with a very low level of impurities. ETBE is an ether resulting from a catalytic reaction between bioethanol and isobutylene, just as MTBE is processed from methanol. MTBE is the most commonly known fuel ether, used in European gasoline since 1973 and still generally used throughout the world to improve gasoline.

Bioethanol can also be the alcohol used for the production of biodiesel, thus supplying ethyl esters instead of methyl esters. The other constituent is vegetable oil from rape, sunflower, soybeans or other sources.

5.6   Combustion

Direct combustion of biomass is the established technology for converting biomass to heat at commercial scales. Hot combustion gases are produced when the solid biomass is burned under controlled conditions. The gases are often used directly for product drying, but more commonly the hot gases are channelled through a heat exchanger to produce hot air, hot water or steam.

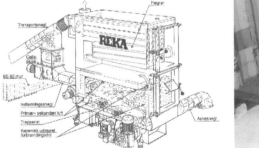 

Figure 69:   Boiler for hot water supply with automatic pellet feeding system by REKA (left); Stirling engine of Indian origin with external combustion using briquettes (right).

At present, the generation of electricity from biomass uses the combustion systems. This process consists of creating steam and then using the steam to power an engine or turbine for generating electricity. Even though the production of steam by combustion of biomass is efficient, the conversion of steam to electricity is much less efficient.

Where the production of electricity is to be maximized, the steam engine or turbine will exhaust into a vacuum condenser and conversion efficiencies are likely to be in the 5-10% range for plants of less than 1 MWe, 10-20% for plants of 1 to 5MWe and 15-30% for plants of 5 to 25MWe.

5.7   Gasification

Biomass can be converted to synthesis gas (consisting primarily of carbon monoxide, carbon dioxide, and hydrogen) via a high temperature gasification process.

Figure 70:   Wood chips gasifier by Volund for district heating CHP in Harboore, Denmark (left); automatic feeding of wood chips with 40% humidity (right). Photo: Volund.

Hydrogen can be recovered from this syngas, or it can be catalytically converted to methanol. It can also be converted using Fischer-Tropisch catalyst into liquid stream with properties similar to diesel fuels, called Fischer-Tropisch diesel.

Gasification technology has been under intensive development for the last two decades. Large-scale demonstration facilities have been tested and commercial units are in operation worldwide. The problems with the application of gasification have been economic, more than technical. In the past, the product from gasification has been electricity or heat, and the low value of these products in today's market is insufficient to justify the capital and operating costs. However, if gasification is coupled with the production of a higher value liquid fuel, the combination could be a viable alternative energy technology.

After gasification, anaerobic bacteria such as *Clostridium ljungdahlii* are used to convert the CO, CO_2, and H_2 into ethanol. Higher rates are obtained because the process is limited by the transfer of gas into the liquid phase instead of the rate of substrate uptake by the bacteria.

Syngas fermentation technology can be used to produce ethanol from cellulosic wastes with high yields and rates. The process of combined gasification/fermentation has been under development for several years. The feasibility of the technology has been demonstrated, and plans are under way to pilot the technology.

5.8   Pyrolysis

Pyrolysis is the thermal degradation of carbonaceous material in the absence of air or oxygen. Temperatures in the 350-800°C range are most often used. Gas, liquid and solid products (char/coke) are always produced in pyrolysis reactions, but the amounts of each can be influenced by controlling the reaction temperatures and retention time. The output of the desired product can be maximized by careful control of the reaction conditions. Because of this, heat is commonly added to the reaction indirectly.

Industrial processes are the most common use of present pyrolysis. An overall efficiency of 35% by weight can be achieved by maximizing the output of the solid product through the implementation of long reaction times (hours to days) and low temperatures (350°C). Heat for the process in traditional kilns can be produced by burning the gas and liquid by-products.

Fast pyrolysis is a high temperature process in which biomass is rapidly heated in the absence of oxygen. As a result it decomposes to generate mostly vapours and aerosols and some charcoal. After cooling and condensation, a dark brown mobile liquid is formed which has a heating value about half that of conventional fuel oil. While it is related to the traditional pyrolysis processes for making charcoal, fast pyrolysis is an advanced process carefully controlled to give high yields of liquid.

Figure 71:   Two types of industrial plants for bio-oil production from biomass crops and residues

The main product, bio-oil, is obtained in yields of up to 80% wt on dry feed, together with by-products of tar and gas which are used within the process so there are no waste streams. While a wide range of reactor configurations have been operated, fluid beds are the most popular configurations due to their ease of operation and ready scale-up.

5.9   Cogeneration

Cogeneration, also known as Combined Heat and Power or CHP, is the production of electricity and heat in one single process for dual output streams. In conventional electricity generation 35% of the energy potential contained in the fuel is converted on average into electricity, whilst the rest is lost as waste heat. Even the most advanced combined-cycle technologies do not convert more than 55% of fuel into useful energy.

Cogeneration uses both electricity and heat and therefore can achieve a combined efficiency of up to 90%, providing very significant energy savings when compared with the separate production of electricity from conventional power stations and of heat from boilers. With 30- 45% of the energy consumed converted into electricity and the balance for heat, this is the most efficient way to use the fuel for electricity production - and for heat production be even more improved by applying heat pumps. Cogeneration helps save energy costs, improves energy security of supply and create jobs.

Figure 72:   Combined heat and power station with hot water storage tanks for town of 600 inhabitants (left); one of the two 720 kW Jenbacher gas engines (middle); preinsulated heating pipes for district heating (right).

The heat produced by cogeneration can be delivered through various mediums, including warm water (e.g., for space heating and hot water systems), steam or hot air (e.g., for commercial and industrial uses). It is also possible to do trigeneration, the production of electricity, heat and cooling (through an absorption chiller) in one single process. Trigeneration is an attractive option in situations where all three needs exist, such as in production processes with cooling requirements.

Cogeneration schemes are usually sited close to the heat and cooling demand and, ideally, are built to meet this demand as efficiently as possible. Under these conditions more electricity can be sold to the electricity grid or supplied to another customer via the distribution system.

In recent years cogeneration has become an attractive and practical proposition for a wide range of applications. These include the process industries (pharmaceuticals, paper and board, brewing, ceramics, brick, cement, food, textile, minerals, etc.), commercial and public sector buildings (hotels, hospitals, leisure centres, swimming pools, universities, airports, offices, barracks, etc.) and district heating schemes. While in the European Union 9% of the electricity is produced in combination with heat, in Denmark there is a significantly higher share of over 50% as part of a policy to improve overall energy- and cost-efficiency.

Figure 73: Small 9 kW combined heat and power unit with Lister engine fuelled with PPO.

Cogeneration can be based on a wide variety of fuels an individual installations may be designed to accept more than one fuel. While solid, liquid or gaseous fossil fuels dominate currently, cogeneration from biomass fuels is becoming increasingly important. Sometimes, fuels are used that otherwise would constitute waste, e.g., refinery gases, landfill gas, agricultural waste or forest residues. These substances increase the cost-efficiency of cogeneration.

5.10 Wind Energy

A windmill converts the energy in the wind into electrical energy or mechanical energy to pump water or grind cereals.

The most common windmills in operation today generate power from three blade horizontal axis windmills with the nacelle mounted on steel towers that can be cylindric steel plate or lattice towers. This modern windmill concept has emerged since 1977 and has become the industrial standard. The nacelle of horizontal-axis windmills usually includes a gearbox, asynchronous generator, and other supporting mechanical and electrical equipment (see diagram below).

A German manufacturer of windmills has obtained a leading international position by using gearless, multipole synchronous ring generators. All contemporary windmills have advanced control and safety systems with automatic orientation to the actual wind direction. They have a design lifetime of 20 years. Wind turbines are rated by their maximum power output in kilowatts (kW) or megawatts (1,000 kW or MW). For commercial utility-sized projects, the most common windmills sold are in the range of 600 kW to 2 MW – large enough to supply electricity to 600-2,000 modern homes. The newest commercial turbines are rated at 1.5-2.5 megawatts. A typical 600 kW turbine has a rotor diameter of 45 metres and is mounted on a 50 metre or taller concrete or steel tower depending on the local wind resources and topography. The newest (in 2003) megawatt-size types have rotor diameters up to 120 metres.

Figure 74:   Wind farms with megawatt-size windmills by Bonus: Sited in open farmland (left); off-shore installation (right). Photo by V. Kantor

The power generated from a modern windmill is practically related to the square of the wind speed, although theoretically it is related to the cube of the wind speed. This means that a site with twice the wind speed of another in practice will generate four times as much electricity. Consequently, the availability of reliable wind speed data is critical to the feasibility of any wind project.

Data is usually gathered over a period of time using anemometers installed at the prospective site. Normally, one year is the minimum time a site should be monitored. In several countries the wind resource assessment occurs by applying the wind atlas method which secures a predictability of the power production with accuracy better than 10% of what in practice is achievable at the actual site.

Most commercial wind turbines operating today are at sites with average wind speeds higher than five metres/second (m/s) or 18 km/h. A prime wind site will have an annual average wind speed in excess of 7.5 m/s (27 km/h).

Figure 75: Wind farm in landscape with barley fields.

Utility-size commercial wind projects are often constructed as wind farms where dozens turbines, totalling 50 MW or more, are erected at the same site. Wind projects have been successfully built to power a wide range of applications in diverse and often extreme environments. One of the newest applications is to a place wind farms in shallow off-shore areas with water depths of 30 metres or more. At sea, environmental impacts are often lower and the availability of a steady, non-turbulent wind flow allows windmills to operate more efficiently and generate 50% more power compared to good on-shore sites.

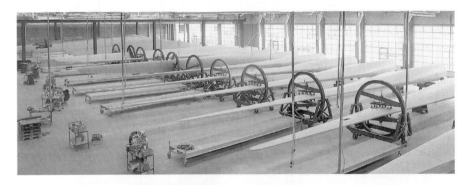

Figure 76: Manufacturing of modern megawatt-size windmills: Factory for 30 meter blades at LM (top). Photo by LM; assembly of Enercon nacelle (left); ring generator fabrication at Enercon (right). Photos by BWE.

Although the wind resource for any site is intermittent, it can, by following weather forecasts, be highly predictable. Thus the output from windmills can be integrated into existing electrical grids with a high degree of dependability. A modern windmill's "capacity factor" (the percentage of time a windmill has delivered nominal power) is in the range of 20-40 percent depending on the wind resources.

Electrical utilities can generally absorb twenty percent of their generating capacity from intermittent sources such as wind without power quality problems. However, in Europe some local utilities that are connected to the national grid receive over 100% of their power needs from local windmills. During periods of strong winds and low power consumption, the windmills may deliver up to 400% of the actual consumption of electricity with the prescribed power quality still being maintained.

Distributed power generation with connection of the windmills to the existing supply network is often being practised with proper calculations of peak values and the corresponding grid reinforcement. Quickly dispatched capacity such as hydro can allow a larger percentage of the overall capacity to come from windmills. The newest variable speed windmills can also help to stabilize grids in remote locations. The feasibility of a wind project, however, can be influenced by access to the electrical grid. The need to install or upgrade high voltage transmission equipment can significantly add to the cost of a wind project. For off-grid and mini-grid applications, the combination of wind/diesel or other sources can provide a greater percentage of overall capacity.

5.10.1 Types of Windmills

Early windmills - less than twenty years ago - were fairly small (50-100 kW, 15-20 m diameter) but there has been a steady growth in size and output power. Several commercial types of windmills in 2003 have ratings over 1.5 MW and machines for the off-shore market have outputs up to 3 MW.

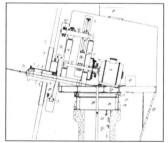

Figure 77:   Two hundred years of wind energy utilisation: Dutch type windmills for grinding (left); roof top mechanical farm windmill from 1910 - 39 for water pumping, threshing, grinding etc. (2nd left); "mother" of modern wind technology, 200 kW Gedser windmill in Denmark (right).

Windmill sizes have increased for two reasons. They are cheaper and they deliver relatively more energy. The energy yield is improved partly because the rotor is located higher from the ground and so intercepts higher velocity winds, and partly because they are slightly more efficient. The higher yields are clearly shown in Figure 78 that shows data from machines in Denmark. The productivity of the 600 kW windmills is around 50% higher than that of the 55 kW windmills. Reliability has improved steadily and most windmill manufacturers now guarantee operational availabilities of 95%.

The majority of the world's windmills have three glass-reinforced polyester blades. The traditional mechanical power train includes a low speed shaft, a speed-up gearbox and an asynchronous generator, either four or six-pole. There are other possibilities, however. Wood-epoxy is an alternative blade material and some machines have two blades.

Towers are usually made of steel and the great majority is of the tubular type. Lattice towers, common in the early days, are now rare, except with smaller windmills in the range 100 kW and below. The foundation is usually a reinforced concrete structure. The tower can also be bolted directly to solid rocks when available.

Variable speed windmills are becoming more common and most generate power using an AC/DC/AC system. Variable speed brings several advantages; it means that the rotor turns more slowly in low winds (which keep noise levels down), it reduces the loads on the rotor, and the power conversion system is usually able to deliver current at any specified power factor. A few manufacturers build direct-drive windmills, without a gearbox. These are usually of the variable speed type with power conditioning equipment.

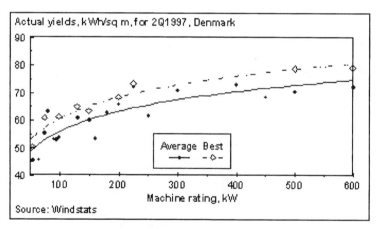

Figure 78: Energy productivity and machine rating.

As the power in the wind increases with the cube of the wind speed, all windmills need to limit the power output in winds of 15 m/s and higher. There are two principal means of accomplishing this: Variable pitch control of the blades or fixed, stall-regulated blades. Pitch-controlled blades are adjusted automatically as wind speeds increase so as to limit the power output once the "rated power" is reached. A reasonably steady output can be achieved, subject to the control system response. Stall-controlled rotors have fixed blades that gradually stall as the wind speed increases, thus limiting the power by passive means.

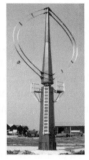

Figure 79:   Wind power pioneering examples: First contemporary windmill in South America was 75 kW unit at Fernando do Noronha, Brazil, 1992, by Folkecenter and Eolica (left); Vestas darrieus-type windmill from 1978 (middle); Tvind 2 MW 3-bladed down-wind windmill from 1977 with advanced fibre-glass blade technology (right). Photo by V. Kantor 2002.

The simplicity of stall control dispenses with the necessity for a pitch control mechanism, but it is rarely possible to achieve constant power as wind speeds rise. Once peak output is reached the power tends to fall off with increasing wind speed, and so the energy capture may be less than that of a pitch-controlled windmill. The merits of the two designs are finely balanced, which accounts for the roughly equal numbers of these windmill types.

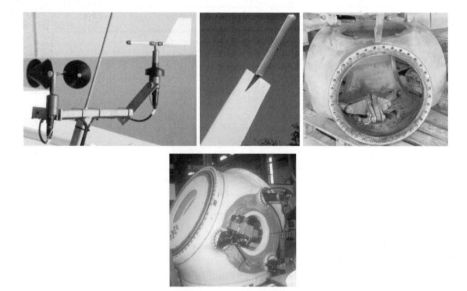

Figure 80:   Wind speed anemometer and wind orientation vane on top of nacelle (left); tip-brake for over-speed protection (2nd left); hub for stall regulation (right); hub for pitch-control (below).

5.10.2 Energy Yield

Energy yieldsdo not in practice increase with the cube of the wind speed, mainly because energy is discarded once the rated wind speed is reached. To illustrate a typical power curve and the concept of rated output, Figure 82 shows a typical performance curve for a 1.65 MW windmill. Most windmills start to generate at a similar speed - around 3 to 5 m/s - and shut down in very high winds, generally around 22 to 25 m/s.

Annual energy production from the windmill whose performance is charted in Figure 82 is around 1 500 MWh at a site where the annual average wind speed is 5 m/s, 3,700 MWh at 7 m/s and 4,800 MWh at 8 m/s. Wind speeds around 5 m/s can be found, typically, away from the coastal zones in all continents, but developers generally aim to find higher wind speeds.

Levels around 7 m/s are to be found in many coastal regions and sites around the North Sea; higher levels are to be found on many of the Greek Islands, in the Californian passes - the scene of many early wind developments - and on upland and coastal sites in the Caribbean, Ireland, Scotland, the United Kingdom, Spain, New Zealand and Antarctica.

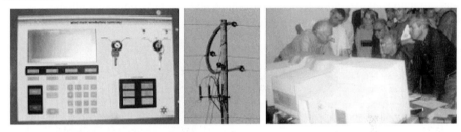

Figure 81: Computerized control panel for modern windmill (left); interfacing between windmill transformer and 10 kilovolt grid (middle); local power utility controlling distributed generation from 220 farmer owned windmills (right).

Wind speed is the primary determinant of electricity cost, on account of the way it influences the energy yield. Roughly speaking, developments on sites with wind speeds of 8 m/s will yield electricity at one third of the cost compared to a 5 m/s site. Off-shore wind speeds are generally higher than those on-shore.

Off-shore wind farmshave been completed, or are being planned in Denmark, Sweden, Germany, the United Kingdom, Ireland and elsewhere. The first off-shore wind farm in the North Sea was established in 2002 and consists of 80 windmills with a total capacity of 160 MW. The annual electricity production is 600 GWh covering the power consumption of 150,000 modern families.

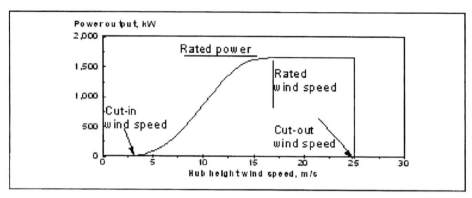

Figure 82: Power curve for a 1.65 MW windmill

Off-shore wind energy is attractive in locations such as Denmark and the Netherlands where pressure on land is acute and windy hill top sites are limited. In these areas off-shore winds may be 0.5 to 1 m/s higher than on-shore, depending on the distance. The higher wind speeds do not usually compensate for the higher construction costs but the chief attractions of off-shore are a large resource and low environmental impact.

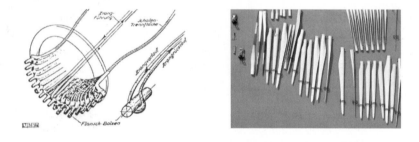

Figure 83: Windmill blade technology: Hütter fibre-glass-to-steel assembly method first implemented by Tvind and Økær (left); selection of blades (right); transportation of 38 meter LM blade (bottom). Photos by LM.

5.10.3 Small Windmills

There is no precise definition of "small", but it usually applies to windmills under about 30 kW in output. In developing countries small windmills are used for a wide range of rural energy applications, and there are many "off-grid" applications in the developed world as well - such as power for navigation beacons. Since most are not connected to a grid, many use DC generators and run at variable speeds. A typical 100 W battery-charging windmill has a shipping weight of only 15 kg.

Figure 84: Proven 2,5 kW PMG windmill for battery charging (left); Calorius 5 kW windmill with water brake for heat supply (middle); Whisper 0.9 kw wind battery charger with carbon fibre blades (right).

A new unexpected application for small windmills has emerged especially in USA in the name of "home power". Instead of buying power from the grid, individuals prefer to produce their own electricity from photovoltaics, small windmills or a combination of both which saves battery storage capacity. In this way house owners are taking control of the type of the energy source. It is assumed that 100,000 families in the USA are self-reliant in terms of energy supply using renewable energy - and that has formed the background for a significant new market for home power equipment.

Figure 85: Gaia 5,5 kW grid connected windmill during installation (left); testing combined photovoltaic and windpower. Solar panels serve as roof for bus stop shelter (right).

There are several manufacturers of windmills ranging from 100 Watt up to 10 kW some of them using advanced carbon fibre blades, and permanent magnet generators (PMG).

This type of equipment provides great opportunities for paving the way for wind energy in the many un-served areas of the world. In these regions the electricity from small windmills will be higher than from large windmills - but generally lower compared to photovoltaics wherever wind resources are limited. A niche market, where wind turbines often come into their own as the costs of energy from conventional sources can be very high, is in cold climates. Wind turbines may be found in both polar regions and in northern Canada, Alaska, Finland and elsewhere. To illustrate the point about economic viability, data from the U.S. Office of Technology Assessment quotes typical costs of energy at 10 kW capacity in remote areas:

Micro-Hydro	~ EUR 0.21/kWh
Windmill	~ EUR 0.48/kWh
Diesel	~ EUR 0.80/kWh
Grid Extension	~ EUR 1.02/kWh

5.10.4 Environmental Aspects

No energy source is free of environmental effects. As the renewable energy sources make use of energy in forms that are diffuse, where larger structures, or greater land use tends to be required - attention may be focused on the visual effects. In the case of wind energy, there is also continues a discussion of the effects of noise irrespective that modern windmills, in some places operate at extremely low noise levels of 40 dB(A) near residential areas.

Possible disturbance to wildlife focuses especially at birds where killings are below 0.1% -compared to animal fatalities in road traffic and collisions with power lines. It must be remembered, however, that the main reasons for developing the renewable sources are environmental in order to reduce emissions of greenhouse gases and the substitution of atomic energy and fossil fuels.

5.10.5 Economic Aspects

World wind energy capacity has been doubling every three years during the last decade and growth rates in the last two years have been even faster, as shown in Figure 86. In 2002 almost 7,000 MW new capacity of windpower was installed worldwide bringing the total capacity to over 31,000 MW by end of 2002 delivering 0.4% of world electricity demand. This is four times more than the 7,600 MW total of 1997 and indicates wind energy is the strongest new energy growth source.

The countries with the most wind power capacity are Germany - by far the largest, with 12.000 MW - followed by Spain, the United States, Denmark and India. In Germany wind energy covers almost 6% of total power demand and in Denmark 20% (2002). With stronger political commitments worldwide, installed wind energy could attain an estimated 230,000 MW by 2010 and 1.2 million MW by 2020.

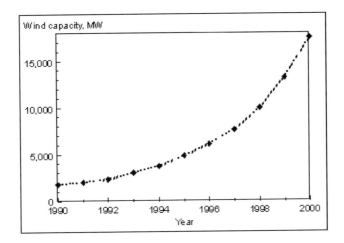

Figure 86: Growth of world wind capacity

The attractions of wind as a source of electricity which produces minimal quantities of greenhouse gases has led to ambitious targets for wind energy in many parts of the world. More recently, there have been several developments of off-shore wind installations and many more are planned. Although off-shore wind-generated electricity is generally more expensive than on-shore, the resource is very large and there are few environmental impacts. While wind energy is generally developed in the industrialised world for environmental reasons, it has attractions in the developing world as it can be installed quickly in areas where electricity is urgently needed. In many instances it may be a cost-effective solution if fossil fuel sources are not readily available. In addition there are many applications for wind energy in remote regions, worldwide, either for supplementing diesel power (which tends to be expensive) or for supplying farms, homes and other installations on an individual basis.

Figure 87: Market place for second hand windmills in Thy, Denmark (left); 75 kW second hand WindMatic windmill used for training of students at St. Petersburg State Technical University (right).

Wind energy is, at good wind sites, generally cost-competitive with the conventional sources of electricity generation. However, the pattern of development has been largely dependent on the legal frameworks provided by national governments to compensate for the negative conditions wind energy will find in competition with conventional fuels that causes environmental damages and also receive several types of direct and indirect subsidies.

Wind energy costs have declined steadily and a typical installed cost for on-shore wind farms is now around EUR 1 000/kW installed, and for off-shore around EUR 1 600/kW. The corresponding electricity costs vary, partly due to wind speed variations, financial terms and partly due to differing institutional frameworks. Wind prices are converging with those from the thermal sources but it is not easy to make objective comparisons, as there are few places where totally level playing fields exist.

Two examples may be given. Until recently, the UK in the 1990's operated a competitive tender market for renewable energy sources with guaranteed payments for 15 years. Vigorous competition drove prices down rapidly and the prices realised in the last round of the Non-Fossil Fuel Obligation may be compared with prices for new gas and coal-fired plants. These comparisons, shown in Figure 88, show that wind prices are very similar to those for coal-fired plants and only a little more than those of gas-fired plants. However, many cost-efficient UK project plans did not materialize due to local citizens' protests.

The second set of comparisons has been drawn from two USA sources: a Department of Energy projection for 2005 and a recent analysis for the State of Oregon in 2000. This comparison shows a bigger gap between wind and gas although wind is significantly cheaper than nuclear. Other USA data suggest that wind prices down to around 4 US cents/kWh can be realised in some areas.

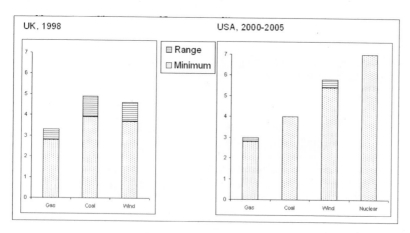

Figure 88: Electricity prices (in US cents/kWh) for wind and the thermal sources, UK and USA

5.11 Solar Energy

5.11.1 Solar Technologies

Solar technologies – some primitive, some more advanced – have been used in all ages and in every corner of the world, but the invention and development of modern solar technologies goes back only forty or fifty years. By now the world has seen numerous practical demonstrations that sophisticated solar-powered facilities can be built and operated successfully as part of energy systems ranging from the scale of an individual home, to a large industrial or commercial complex, or even a whole city, an island or rural area. In principle, every roof or façade of buildings can in the future become a solar power plant. Solar energy is not only to be found in parts of the world with intensive solar radiation. Photovoltaics are being used even for supply of electricity for telecommunications in remote mountains on Greenland north of the Polar Circle.

As early as the 1980's, a 354 MW solar power plant was built in the Mojave desert, in California. Here the heat, contained in solar rays concentrated by reflecting troughs and raised to 400°C, produces steam that runs a conventional power generator. When the sun is not shining, the plant switches to natural gas. The latest generation of this type of plant incorporates new engineering solutions and new scientific principles such as non-imaging optics, which makes it possible to build much more efficient concentrators at lower costs. These developments open new prospects for the technology in the sunniest parts of the world.

A solar technology that has already had a great impact on our lives is photovoltaics, not in terms of the amount of electricity it produces (in 2002 2,400 MW were installed), but because of the fact that photovoltaic cells – working silently, not polluting – can generate electricity wherever the sun shines, even in places where no other form of electricity can be obtained.

The technology has been around since the 1950's, but the effect on our lives is not widely known. As the American solar-technology historian John Perlin observes, it was the determining factor in a whole series of otherwise unthinkable developments. For instance, photovoltaic cells generate the power that runs space satellites. Without telecommunications satellites, many of our now-routine activities – from watching internationally broadcast entertainment to using cell phones – would still be in the realm of science fiction. And space exploration and research too might still be science fiction.

On earth, photovoltaic technology is used to produce electricity in areas where power lines do not reach. In the developing countries, it is significantly improving living conditions in rural areas. Thanks to its flexibility, it can be incorporated in packages of energy services and thus offer unique opportunities to improve rural health care, education, communication, agriculture, lighting and water supply.

*Figure 89:   Integration of PV in buildings: Electronic "curtains" in window panes (left); solar roof panels
             in residential building (middle); semi-transparent thin film PV panels in glass corridor (right)
             Examples developed by Folkecenter.*

In the industrialised countries, programmes that provide incentives for the incorporation
of photovoltaic systems in building roofs and walls have tallied up thousands of
completed projects in Japan, Europe and USA. Japan has demonstrated the use of
photovoltaics in very successful programmes while the German 100,000 roofs
programme in 2003 is close to being implemented. In both countries the way has been
paved for many new jobs in industries with a leading global position. This has raised the
awareness in the population of energy issues and protected the environment by
substituting the combustion of fossil fuels.

Annual worldwide sales of photovoltaic systems are growing by around 30-40% and
stands at about five billion EUR in 2003. The leading manufacturing countries are Japan
and Germany with USA coming next. Large corporations within the electronic
equipment industry with worldwide distribution are building or planning solar cell
factories with an annual capacity of 50 MW or more. In 1996 global production of
photovoltaic modules was 65 MW which in 2002 had grown to 400 MW illustrating the
rapid growth. The potential for further growth is very high and similar to wind energy
irrespective of the relatively high costs of solar energy.

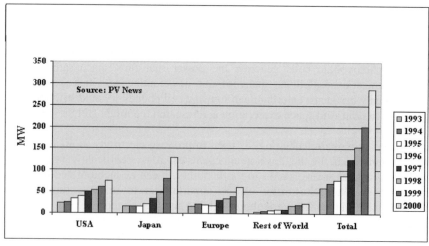

Figure 90:   World cell/module shipments 1993-2000 (consumer and commercial).

Photovoltaic (PV) systems are a reliable, renewable, environmentally safe, and increasingly cost-effective technology for generating electricity for a wide range of applications vital to households and communities in the developing world and the industrialized countries as well. PV modules can power small systems, such as simple fluorescent lighting and communication for individual homes, and they can be linked to run larger-scale water pumps, communications equipment for schools, and appliances that benefit communities as a whole.

For developing countries, where a substantial proportion of the rural population in un-served areas typically lacks access to the most basic energy services and may not be reached by grid based power for decades if ever, PV can help provide much-needed services. Even the smallest solar home system can yield a level of illumination equivalent to 60 kerosene lamps.

This kind of energy service is a quantum advance in quality of life that reinforces developmental and anti-poverty objectives - for example, by allowing children to do homework in the evenings or making it possible for families to listen to a radio or watch a small television. Similarly, a few PV panels can power a refrigerator that keeps vaccines cold in a health clinic.

PV systems are also environmentally and technologically well suited to developing countries. They produce no emissions and conserve local resources. As costs drop and environmental externalities are factored in, PV should become a major factor in global energy supply in the near future, both in grid-connected and stand-alone applications.

At present, PV systems are especially useful in developing countries for individual and small community applications. They are modular, so panels, batteries, and other equipment can be installed in single or multiple units. They also can be easily transported, rapidly installed, and conveniently expanded, often by local technicians and entrepreneurs, as power requirements grow. Finally, because they lack moving parts, PV systems are low-maintenance, long-service technologies.

5.11.2 Photovoltaics (PV)

Photovoltaics (PV) convert sunlight directly into electricity. Photons in sunlight interact with the outermost electrons of an atom. Photons striking the atoms of a semiconducting solar cell free its electrons, creating an electric current. The Photovoltaic effect was first discovered in the 19th century, and was used by Bell Labs in 1954 to develop the first PV solar cell.

PV found its first applications in space, providing electricity to satellites. These early PV cells were produced in small quantities from exotic materials. While early cells were inefficient, converting less than 1% of the incident sunlight into electricity, they quickly increased to 6% when researchers experimented with crystalline silicon, the principal component of sand. Current conversion efficiencies have surpassed 30% in the laboratory, and 15% in large-scale production.

Two main types of silicon cells vie for market share: crystalline and thin-film. Crystalline silicon cells are produced by slowly extracting large crystals from a liquid silicon bath. These crystals are sliced into 1/100th-of-an-inch thick slices, or "wafers", which are processed into solar cells that are then connected and laminated into solar "modules." While this production process yields highly efficient (10-15%) cells, the production process is rather expensive.

Thin-film silicon cells are produced by depositing vaporized silicon directly onto a glass or stainless steel substrate. While the efficiencies achieved are lower than with crystalline silicon, the production process is less expensive. Modules from crystalline cells have a lifetime of over twenty years. Thin-film modules will last at least ten years. Other PV technologies, such as Gallium-Arsenide or Cadmium Telluride, are also being used. These types are highly efficient, but more expensive at the present time.

Most commercial PV modules are based on wafer-thin slices sawn from mono- or poly-crystalline ingots of high-grade silicon. Monocrystalline ingots are grown in a "batch" process. Although the method is slow and energy intensive, it produces a cell with good conversion efficiencies (typically 12 to 17 percent). Polycrystalline PV materials, formed by the less arduous method of casting ingots from many small silicon crystals, generally have slightly lower conversion efficiencies. Both types, if encapsulated and given proper care, do not degrade in performance. Figure 91 shows how crystalline silicon is produced into a PV module. Although silicon is an abundant natural resource, supplies of high-grade raw material at a suitable price may not be sufficient to enable crystalline modules alone to meet the anticipated growth in the PV market. Increasingly, so-called thin-film materials (produced by depositing the PV materials as a fine layer onto a base substrate) are being seen as preferable for PV modules in the long term because they use fewer raw materials and are better suited to cost-effective mass production processes.

Among the thin-film materials, amorphous silicon is now the most commonly used. It could play a significant role in low-power modules (less than 20Wp), or in PV facades for buildings, where its comparatively low efficiency (typically 4 to 6 percent) does not present a significant problem. Moreover, stacking layers of amorphous cells yields much better conversion efficiencies. These configurations show promise for future power applications, but long-term field tests are needed to confirm their performance and reliability.

Several other thin-film technologies, including cadmium telluride and copper indium diselenide, are now approaching commercialisation. These could deliver reductions of cost in mass production similar to amorphous silicon and should have higher stable conversion efficiencies.

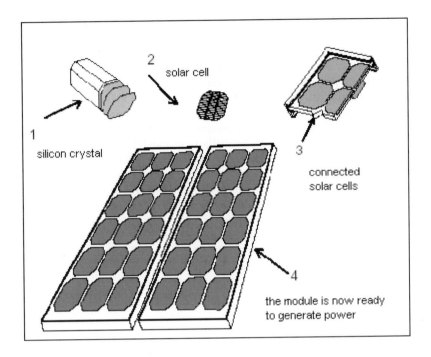

Figure 91:   Production of a Crystalline Silicon PV Module

Components

PV systems may comprise some or all of the following basic components (see Figure 92):

- PV module or array of modules and accompanying support structures. PV power-generating modules can be expected to operate for 20 years or more, assuming they are certified to International Electro technical Commission Standards, and they should not require much more maintenance than occasional cleaning to remove deposits of dirt.
- Power storage (usually provided by batteries). Batteries are required for most off-grid applications except water pumping. They are currently the weak link in the PV system. The inexpensive automotive batteries commonly used in rural areas of developing countries do not operate well under the cycles of deep discharging and recharging required by PV systems and typically must be replaced every two to three years. However, they are widely used often in the form of truck batteries because they are commercially available everywhere. Batteries that are better suited to PV applications and that last longer - for example, tubular plate stationary lead-acid batteries - are being manufactured, but they are expensive and not widely available.
- Power-conditioning equipment (includes Charge controllers, inverters and control and protection equipment). Electricity generated by PV is in the form

of direct current (DC), usually at 12 volts. This direct current is often converted by an inverter into the alternating current (AC) required by many appliances. Inverters themselves can cause energy losses in the conversion process and in standby mode. String inverters, a recent innovation, are becoming more popular. AC modules incorporating mini-inverters are another new development but are not yet in wide use (where PV systems are being integrated with grid power, such modules can be plugged directly into the main grid system, which simplifies connection). The need for an inverter can be avoided by using DC appliances, particularly in standalone applications, where the PV system is not connected to a main electricity grid. In un-served areas, especially in the developing countries, where hundreds of millions of people are living, 24 or 48 Volts are frequently used. Two 60 Watt PV modules, a charge controller and two truck batteries with a total cost of approximately 1,000 EUR are sufficient for the supply of electricity of a village school and provide the opportunity for adult education in the evenings. However, DC products may be difficult to obtain. The best system to use thus will depend heavily on the local appliance markets.

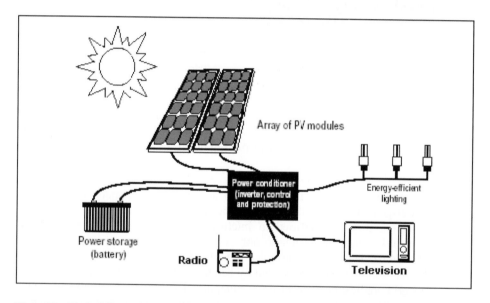

Figure 92: Typical Components of a PV System

PV is measured in units of "peak watts"(Wp). A peak watt figure refers to the power output of the module under "peak sun" conditions, considered to be 1000 Watts per square meter. "Sun hours," or "insolation," refer to how many hours of peak sun, on average, exist in different countries. North America averages 3 to 4 peak sun hours per day in summer while equatorial regions can reach above 6 peak sunlight hours.

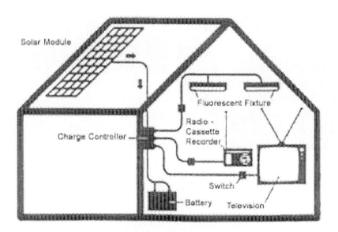

Figure 93: Solar Home System for power supply

Photovoltaic power systems

The broad field of photovoltaic (PV) systems is divided into major three categories (IEA, 1999)

- *Service applications*, which include energy services in integrated applications such as telecommunications, water pumping, remote sensing, medical equipment, parking meters, bus stops, etc.
- *Remote buildings*, such as farms, houses, hotels, education centres
- *Island systems;* these larger PV-diesel systems are mostly located on islands, but also in remote villages, farms or commercial premises.

Under remote buildings, a sub-category of very basic PV systems can be distinguished, mostly referred to as:

- *Solar Home Systems*, basic PV systems, with generally around 50 Wp (generally not more than 200 Wp), that provide just enough energy for a few PL bulbs and some limited use of radio or TV. They are mostly applied in electricification of rural areas in developing countries and leisure cottages.
- *Stand-alone Photovoltaics* (SAPV), also referred to as home power or autonomous systems: applications powered by photovoltaics that operate independently of any electricity grid. The PV generator runs the application and recharges storage batteries, which in turn power the application when there is not enough insolation. If there is a surplus of solar power and the batteries are fully charged, this power is lost.
- *Grid-connected Photvoltaics*: PV generators directly connected to the electricity grid through an inverter. The system can supply any surplus of energy to the grid, or extract energy if the PV generator does not meet the demand.

Figure 94: Installation of 2x60 watt PV modules in village school in Niamala, Mali, by Mali Folkecenter.

Grid-based and grid-connected PV in the industrial regions

The future of PV in the industrialized countries can have a direct bearing on the relevance of PV for developing countries, because innovation, research and development, and economies of scale in production are likely to be promoted by a vigorous market in the developed world. Most people in the industrial countries already have access to grid-based electricity.

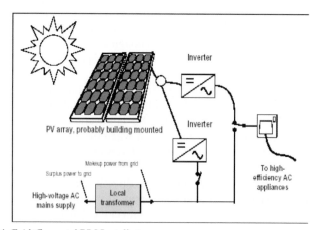

Figure 95: A Grid-Connected PV Installation

Figure 96: Inverters for adapting DC power from photovoltaics to the AC grid. By Fronius (left) and Sunny Boy (right).

Hence, if PV markets are to grow in the industrial countries, PV must be competitive with grid-based power, or it must offer users some unique advantages. Protection of the environment, combined with the decreasing costs and increasing efficiencies of PV, already appears to be leading individuals, institutions, and utilities toward including PV in their energy supply mix.

Rooftop and facade installations
Incorporation of PV into rooftops and facades of buildings is likely to be the main application for PV in many industrial countries, particularly because it avoids or minimizes many costs of traditional power generation, such as purchase of land and building components and transmission and distribution costs. Japan, Germany, Switzerland and the Netherlands are progressing along this path.

The total rooftop generating potential of the OECD countries is estimated at 1,100 GWp (1990); this would be sufficient to meet between 14 and 19 percent of the OECD's estimated 6.8 million GWh annual electricity consumption. In the European Union the Green Book and the preceding White Book have focused at the need of independence of imported fossil fuels and the creation of future-oriented industries through deployment of photovoltaics and other renewable energy forms.

PV Generating Plant
Use of PV in medium- to large-scale generating plant will also be an important application. The United States and Italy are the world leaders in development of large-scale PV systems. Other countries intend to develop "redundant" land alongside motorways and railways, for example by incorporating PV into sound barriers. PV concentrators that use large reflectors to concentrate sunlight onto small cells, saving on PV materials and lowering costs, are also being developed for large-scale PV generating plants.

Figure 97:   Installation of solar panels for lighting of market place in Zambala village in Mali.

PV in the Developing World

With due care and attention, PV offers a reliable and environmentally safe way of providing vital services in rural areas in many parts of the developing world:

- Community lighting. PV-powered lighting, with battery storage, is used for community buildings such as schools and health centres to enable education and income-generating activities to continue after dark.
- Household appliances. Solar home systems can provide power for domestic lighting and other DC appliances such as TVs, radios, and sewing machines.
- Refrigeration. In rural areas, electricity provided by PV modules is particularly valuable for running refrigerators that preserve vaccines, blood, and other consumables vital to community health care programmes.
- Battery charging. PV modules can be used to recharge batteries that power electric appliances ranging from torches (flashlights) and radios to television and lights.
- Water pumping. PV is commonly used to supply water to villages, to irrigate land, and to water livestock.

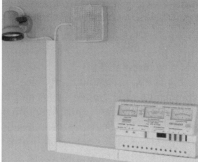

Figure 98:   Installation of medicine refrigerator powered by PV for policlinic in Tabakoro village in Mali (left); charge controller for stand-alone PV installation (right).

Stand-Alone Applications

Measured by averaged unit energy costs using traditional accounting mechanisms without considering the environmental externalities - PV-generated electricity cannot yet compete with efficient conventional central generating plant feeding a main grid.

However, for isolated buildings and small, dispersed communities, stand-alone PV systems - generally with battery storage - are a cost-effective power supply option. More than 25 percent of annual PV sales are currently used in remote professional applications, such as communications repeater stations and lighthouses, which require a small and often intermittent but reliable power supply. A further 15 percent serve the "leisure" market, mainly in developed countries, providing power to remote holiday homes or for camping and boating.

Mini-Grid and Hybrid Systems

But PV power is not necessarily confined to individual installations in niche markets. For communities with many potential connections, a *mini-grid* or *mini-utility* could be appropriate. Such an entity might use PV as the sole power source, or it might combine PV in a hybrid system with a diesel generator or with another renewable source of energy such as wind or mini-hydro that complements the availability of solar power. In the United States and Australia, in particular, PV Remote Area Power Supply (RAPS) is increasing in importance as an alternative or complement to diesel generators for isolated households and remote community power supplies.

Obstacles

Nevertheless, some serious non-technological barriers impede the widespread deployment of small, comparatively inexpensive PV systems. To begin with, governments, major international finance organizations, and development banks are geared toward financing traditional thermal or hydropower projects with a single low interest and long term lending point similar to the investment conditions that are found within the conventional electricity sector. Other obstacles appear on the individual or local level, such as financing purchases by widely dispersed households and meeting the corresponding sales and service demand. These issues - and some innovative credit and policy approaches toward financing first costs - will be discussed in subsequent notes.

Promotion and Economics

Governments also may be encouraged to promote development of PV systems both grid-connected and stand-alone, because of the high costs of constructing large centralized power stations and extending grid electricity. Moreover, traditional power stations often rely on imported fossil fuels, which can lead to increased balance of payments deficits and supply insecurity as well as environmental costs, whether accounted or not. Countries seeking to extend the benefits of electrification may consider stand-alone systems, mini-grids, and ultimately, incorporation of PV into centralized power systems.

Until comparatively recently, the price of PV systems was prohibitively high for most terrestrial applications. With gains in production experience and improved manufacturing methods, though, the costs of PV production have come down significantly from more than EUR 30/Wp 20 years ago to about EUR 4/Wp in 2002. This has allowed substantial reductions in the overall price of PV systems and has transformed PV into a cost-effective energy supply option for a huge range of terrestrial applications. The total costs of a PV system depend on many factors aside from the basic hardware. However, the costs of the PV array are a significant factor and will typically constitute 30 to 50 percent of the total capital outlay, with the BOS components contributing a similar amount. Costs for typical configurations are sketched below:

A small solar home system to power two or three fluorescent tubes, a radio, and a black-and-white television cost EUR 350 to EUR 700. This system typically requires a 50Wp module and support structure, charge controller, battery and wiring.

- A PV-powered vaccine refrigerator - cost, EUR 2,000 to EUR 5,000. This system might require a 200Wp array and would have a well-insulated case, a high-performance battery, along with cables and controllers.
- A roof-mounted community or school PV system running lights, electrical equipment, and so on - cost, EUR 10,000. The cost of the electricity produced by this 1.5kWp (i.e., 1,500 Wp) system would depend on the overall system efficiency, the resource availability, the lifetime of the system and the assumed discount rate, but unit costs are typically on the order of EUR 0.35 to EUR 0.65/kWh. With increased efficiencies of scale in production of modules and other components of the system, this figure could fall to around EUR 0.10/kWh. Although this figure is on the high side of costs for grid-based power, it would be of great value to people in off-grid areas and would represent a substantial social saving in terms of grid extension and traditional fuel costs. By using a 10 watt low energy light tube the cost per hour will be affordable to most rural families whereas it would be prohibitive for operation of refrigerator and air-conditioning.

The cumulative "first costs" - that is, the basic outlay for buying and setting up a PV system - may well be beyond the purchasing power of many consumers in developing countries. Several factors may mitigate or exacerbate the impact, suggesting the need for knowledgeable policy interventions.

- People in developing countries sometimes choose to acquire PV systems piece-meal. Householders may start perhaps with a battery that they use to run a light or a radio and then add a PV module and the other components as they recognize their advantages and can afford them. The costs for the fully engineered systems shown above thus may not tell the complete story of the market.
- Most PV system manufacturers will offer a discount for bulk purchases. This can be beneficial for cutting the costs of establishing a new solar program in a region, but it could force out small entrepreneurs in regions where PV markets already exist. Such bulk order arrangements also tend to apply solely to supply and delivery of components and do not address the critical issue of after-sale service.
- Components imported from overseas may be subject to an import levy. India currently has a 50 percent import duty on PV modules, whereas Indonesia has none.
- Costs of systems purchased through local distributors are pushed up by handling mark-ups and local sales taxes. These charges typically amount to about 15 percent of the total cost, but they can be much more.

Like many renewable energy technologies, PV systems have high initial capital costs but very low running costs. These high capital costs can make PV appear unattractive in simple payback terms. However, life-cycle costing, which accounts for all fuel and component replacement costs incurred over the life of the system, puts PV in a far more favourable light compared with traditional energy sources that have low outlays but significantly greater operating costs. To make cost comparison of solar power realistic financial conditions used in the conventional power should be applied being favoured by low interest loans with 20-40 years duration.

There are more than one million photovoltaic systems operating worldwide (2002) in applications ranging from individual consumer products and small-scale stand-alone units for rural use (for example, solar home systems) to grid-connected roof-top systems and large central-grid power stations. In the period 1995-2000, more than 20,000 solar pumps were installed globally.

Typical system size varies from 50 watt (W) to 1 kilowatt (kW) for stand-alone systems with battery storage and small water pumping systems; from 500 W to 5 kW for roof-top grid connected systems and larger water pumping systems; and from 10 kW to megawatts for grid-connected ground-based systems and larger building-integrated systems.

Photovoltaic devices are solid-state devices with no moving parts and a demonstrated record of reliability. PV modules may operate for 30 years and are usually sold with 10-30 year manufacturer warranties that are significantly longer than for all normal appliances. Although PV modules themselves require little maintenance, other components, particularly storage batteries, generally require maintenance. To obtain an optimal performance and reliability of solar power systems, training of local solar electricians should be given special attention.

To accurately assess the value of electricity from a PV system it is necessary to compare the cost of the PV system to the minimum cost of providing the *same energy service* by an equivalent alternative. This is particularly relevant for stand-alone systems in remote areas where the temptation is to simply compare the capital costs between PV and other energy supply options. A more accurate approach is a life cycle cost analysis of a fossil fuel generation system that includes fuel, maintenance, depreciation, interest and other expenses. PV systems generally have a high capital cost but a low running cost, as the "fuel" is sunlight that is freely available everywhere.

For rural villages in developing countries, PV technology offers obvious advantages an immediate, direct and safe alternative to kerosene lamps and diesel generators often using imported fossil fuel. In such countries a solar home system to power lights and small appliances can be purchased for as little as EUR 350 and may be much cheaper than a grid extension or diesel generator.

The Industry and Market Trends
The PV market grew by an average of 30 percent annually from 1995-2002. In 2002, approximately 400 MW of PV modules were sold for a total revenue exceeding EUR 2

billion. The total installed capacity worldwide is over 1,500 MW, with an average cost of approximately EUR 4 per watt. Costs are typically falling by 20 percent for each doubling of cumulative sales. However, for a number of years the PV sector has been a seller's market. Further strong price reductions may be expected in future due to increased mass production and distribution similar to the experiences from other types of electronic equipment.

At least 30 firms worldwide fabricate PV cells and many more assemble these cells into modules. The top ten cell manufacturers control more than 85% of world shipments. Increasing mass production of PV technology continues to reduce costs in line with the classic "learning curve" for new technologies. Since 1975, PV costs have been reduced by 20 percent for each doubling of cumulative sales.

The earlier general consensus that thin-film technologies offer the best long-term prospects for very low production cost has later come into question due to problems of performance and durability. However, crystalline technology for which most new factories are being built still has large potential for cost reduction through manufacturing and commercialisation economies-of-scale and technological improvements. Research and development is aimed at improving both the cell and module efficiencies and reducing the cost for the necessary components, that currently make up half the cost of the system.

The Market for PV
A common misconception about PV systems is that they are suitable only in hot, sunny climates. In fact, the ideal PV generating conditions are cold, bright, sunny days (modules must be prevented from overheating by ensuring good air flow around them, since they obviously should not be shaded). Moreover, PV uses diffuse solar radiation as well as direct sunlight, so areas with perfectly cloudless days are not required. The specific PV generation potential for a site depends basically on the average intensity of the solar energy received there over a year. Detailed local data can be used, when available, to judge whether the average insolation is economic for PV, but for most purposes, global insolation data are available and perfectly adequate.

Health and environmental considerations
Displacing conventional technologies with PV systems can bring various positive effects that are difficult to quantify in direct financial terms. For example, PV provides distinctly better quality of lighting to homes compared with kerosene wick lamps, which are also smoky.

PV-powered electric lighting can enable educational and income- generating activities to continue after dark with reduced risk of fire and without noxious combustion fumes. The World Health Organization has noted that PV offers more reliable refrigeration than other power supply options. This has resulted in increased efficacy of stored vaccines, which in turn has helped to reduce mortality rates.

Common PV modules pose few environmental problems. They produce electricity silently and do not emit any harmful gases during operation. The basic photovoltaic material for most common modules, silicon, is quite benign. Some potential hazards are associated, however, with production of some of the more exotic thin-film technologies. The two most promising options, cadmium telluride and copper indium diselenide, incorporate small quantities of cadmium sulfide, which poses potential risks during manufacture of modules. The well-established protocols for handling such compounds should minimize these hazards.

Battery disposal
Batteries are a more significant environmental concern. Lead-acid types in particular pose potential health and safety risks, as lead persists in the environment and in organic tissue. The issues of collecting, recycling, and ultimately disposing of PV batteries need to be addressed carefully, though it should be kept in mind that lead-acid batteries are used far more extensively in transport than in energy applications.

Deployment issues
From a performance standpoint, there is no significant reason why PV cannot be more widely used. The hundreds of thousands of solar home systems already in service in countries such as Indonesia and the Dominican Republic and the numerous PV refrigerators and water pumping systems installed throughout Africa have worked well. The main problems are the need to provide training for installers and system maintenance staff and to develop suitable support for PV system sales - such as credit facilities and access to replacement components and spare parts.

5.11.3 Solar Thermal

The sun's energy can be collected directly to create both high temperature steam (greater than 100°C) and low temperature heat (less than 100°C) for use in a variety of heat and power applications. The resulting high temperatures can be used to create steam to either drive electric turbine generators, or to power chemical processes such as the production of hydrogen.

In regions with temperate climate the use of energy in the form of heat is one of the largest items in the energy budget. In Europe, for instance, it accounts for around 50% of total energy consumption: around 630 million toe, of which 383 in low-temperature heat and 247 in medium- and high-temperature heat.

Today, low-temperature (<100°C) thermal solar technologies are reliable and mature for the market. Worldwide, they help to meet heating needs with the installation of several million square metres of solar collectors per year.

These technologies can play a very important role in advanced energy-saving projects, especially in new buildings and structures that require large amounts of hot water, heating and cooling. In some regions of Austria, Greece, Germany and Israel thermal solar installations can be found on a large share of private family houses. Several community thermal solar plants in Denmark have been installed supplying hot water for district heating. The plants are using 12 m^2 modules and one installation can have 5,000 m² or more.

Figure 99:   Solar family water heater, Chile (left); solar storage tank of stainless steel with supplementary heating coils for electricity and central heating (middle); advanced solar modules during test in Sakura, Japan (right).

Low temperature solar thermal systems collect solar radiation to heat air and water for industrial applications including:
- *space heating for homes, offices and greenhouses*
- *domestic and industrial hot water*
- *pool heating*
- *desalinsation*
- *solar cooking, and*
- *crop drying.*

These technologies include *passive* and *active* systems. Passive systems collect energy without the need for pumps or motors, generally through the orientation, materials, and construction of a collector. These properties allow the collector to absorb, store, and use solar radiation. Passive systems are particularly suited to the design of buildings (where the building itself acts as the collector) and thermosiphoning solar hot water systems (explained later).

For new buildings, passive systems generally entail very low or no additional cost because they simply take advantage of the orientation and design of a building to capture and use solar radiation. In colder climates, a passive solar system can reduce heating costs by up to 40 percent while in hotter climates, passive systems can reduce the absorption of solar radiation and thus reduce cooling costs.

Figure 100: Testing and demonstration of solar thermal roof panels (left); high-efficient vacuum tubes for hot water supply (middle); solar thermal control box by Arcon (right).

The most common active systems use pumps to circulate water or another heat absorbing fluid through a solar collector. These collectors are most commonly made of copper tubes bonded to a metal plate, painted black, and encapsulated within an insulated box covered by a glass panel, or "glazing". For pool heating and other applications where the desired temperature is less than 40°C, unglazed synthetic rubber materials are most commonly used.

Figure 101: Examples of thermal solar heating in Vienna, Austria: Hot shower at municipal swimming pool (left); solar heaters for elephant shower bath in Schönbrunn Zoo (middle); solar panels on roof-top of residential building in Central Vienna (right).

For domestic applications, the solar hot water system is a mature technology that can provide hot water to meet a significant portion of the hot water needs in a domestic building. In Europe, a SHS can generally meet between 50–65% of domestic hot water requirements, while in subtropical climates, the percentage can be 80-100 % of needs. A domestic SHS ranges in price from about 500 to EUR 2,500. This is higher than a conventional electric or gas hot water system, but a SHS can pay for the extra cost though energy savings. The payback period depends on many factors but is usually in the order of 4-12 years while the useful life of a SHS usually exceeds 15 years.

The same types of solar collectors used in a domestic SHS can also be used for space heating applications. In Sweden and Denmark, large solar district heating systems have been built that heat water for the demand of the summer months but also with large storages for supply of some of the energy needs in the winter heating season.

High temperature solar thermal systems

These use mirrors and other reflective surfaces to concentrate solar radiation. Parabolic dish systems concentrate solar radiation to a single point to produce temperatures in excess of 1000°C. Line-focus parabolic concentrators focus solar radiation along a single axis to generate temperatures of about 350°C. Central receiver systems use mirrors to focus solar radiation on a central boiler. The resulting high temperatures can be used to create steam to either drive electric turbine generators, or to power chemical processes such as the production of hydrogen.

Solar Thermal Electricity

Solar thermal electricity systems collect direct sunlight in special focusing collectors and convert this into thermal energy, which is then used to generate electricity.

Figure 102: The Solar Two project in California USA uses a field of mirrors to focus solar energy onto a central boiler to produce steam and electricity. (Photo courtesy NREL).

The generators in solar thermal electricity systems are driven by steam turbines in the same way as conventional electricity generation. However, the steam is generated by the sun rather than by consumption of fossil fuels or by nuclear heat. Solar systems use special focusing (rather than flat) reflectors to achieve the steam temperatures of more than 350°C required to operate steam turbines efficiently.

Table 27:   Costs of Solar Thermal Energy Supply

Costs

Domestic Solar Water Heating

Capital costs project:	**EUR 1500 - 3000**
Operating life:	*15 – 40 years*
Payback period:	*4 – 14 years*
Maintenance costs:	*EUR 25 - 30/year*

average figures for home owners
Source: Florida Solar Energy Centre

Solar Thermal Electricity

Capital Cost:	*EUR 2500 – 3500/kW**
Operating life:	*20 years*
Liveliest Cost	*EUR 0.8 – 15/kWh***

* kW = kilowatt   ** kWh = kilowatt-hour

Three main types of generator have been demonstrated, all of which require direct sunlight – solar systems are therefore most suitable for Southern European countries and non-EU countries with high direct solar radiation levels, often arid or semi-arid regions:

- Solar farms use parabolic trough reflectors that focus solar radiation onto a line receiver containing the heat transfer medium in pipes. The medium, often thermal oil is collected and passed through a heat exchanger where steam for the turbines is produced. Temperatures used vary between 350°C and 400°C and system size are typically 30-80MW. To increase operating temperatures and thus efficiency, steam from the solar system may be heated in a final stage by conventional fuels to higher temperatures.

- Solar power towers use one central receiver mounted on top of a tower that is surrounded by a field of heliostats – concentrating mirrors that follow the sun. Reflected light is focused onto the receiver and absorbed by the heat transfer medium, which could be sodium, water, molten salt or air. Temperatures of 500-1000°C can be achieved. System sizes up to 200 MW are possible.

- Parabolic dish systems use parabolic concave mirrors which have a receiver mounted at the focus. These systems achieve the highest temperatures, 600-1200°C, but systems are small, 10-50 kW for a single unit. The main application is decentralised electricity generation. The technology's dependence on direct sunlight for sunlight for efficient operation usually limits load factors to less than 25%. One way of improving the load factor is to design plants as hybrid solar/fossil fuel generators.

The Industry and Market Trends

Generating electricity from high temperature solar thermal devices is already a technical reality. There is a small existing market, and costs are currently cheaper than generating electricity from PV for large grid applications. However, there has been only a small increase in the market for this technology over the past decade, and its long-term future depends on the availability and success of further research and development.

For low temperature applications, the market is diverse, with about 180 manufacturers in Europe and the United States alone. At the end of 1998, about 30 million square metres of solar collectors worldwide were providing domestic and commercial hot water. In Europe, the market has grown by 18 percent per year throughout the 1990s and is expected to reach 15 million m^2 by the year 2003.

In Australia, five percent of domestic water heating comes via SHS technology. The growth if this technology and its related industries, however, depends very much on energy policy. Experience in Australia and the Netherlands shows that supportive regulations can increase production volumes, thereby lowering overall costs.

5.12 Hydropower

Hydropower systems use the energy in flowing water to produce electricity or mechanical energy. Although there are several ways to harness the moving water to produce energy, *run-of-the-river systems*, which do not require large storage reservoirs, are often used for micro hydro, and sometimes for small-scale hydro, projects. For run-of-the-river hydro projects, a portion of a river's water is diverted to a channel, pipeline, or pressurized pipeline *(penstock)* that delivers it to a waterwheel or turbine. The moving water rotates the wheel or turbine, which spins a shaft. The motion of the shaft can be used for mechanical processes, such as pumping water, or it can be used to power an alternator or generator to generate electricity.

5.12.1 Types of Hydropower

Impoundment
An impoundment facility, typically a large hydropower system, uses a dam to store river water in a reservoir. The water may be released either to meet changing electricity needs or to maintain a constant reservoir level.

Diversion
A diversion, sometimes called *run-of-river*, facility channels a portion of a river through a canal or penstock. It may not require the use of a dam.

Pumped Storage
When the demand for electricity is low, a pumped storage facility stores energy by pumping water from a lower reservoir to an upper reservoir. During periods of high electrical demand, the water is released back to the lower reservoir to generate electricity.

Table 28, outlines the categories used to define the power output form hydropower. Micro-hydropower is the small-scale harnessing of energy from falling water; for example, harnessing enough water from a local river to power a small factory or village. This fact sheet will concentrate mainly on micro-hydropower.

Table 28:   Classification of hydropower by size.

Large - hydro	More than 100 MW and usually feeding into a large electricity grid
Medium - hydro	15 – 100 MW – usually feeding a grid
Small - hydro	1 – 15 MW - usually feeding into a grid
Mini - hydro	Above 100 kW, but below 1 MW; either stand alone schemes or more often feeding into a grid
Micro - hydro	Ranging from a few hundred watts for battery charging or food processing applications up to 100 kW; usually providing power for a small community or rural industry in remote areas away from the grid.

Figure 103 shows the main components of a run-of-the-river micro-hydro scheme. This type of scheme requires no water storage but instead diverts some of the water from the river that is channelled along the side of a valley before being 'dropped' into the turbine via a penstock. In this fiigure the turbine drives a generator that provides electricity for a workshop. The transmission line can be extended to a local village to supply domestic power for lighting and other uses.

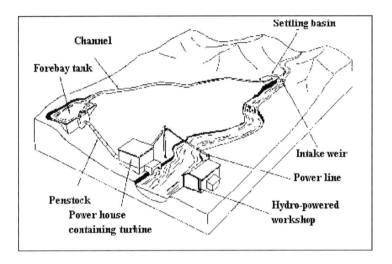

Figure 103: Layout of a typical micro hydro scheme.

There are various other configurations that can be used depending on the topographical and hydrological conditions, but all adopt the same general principle.

To determine the power potential of the water flowing in a river or stream it is necessary to determine both the flow rate of the water and the head through which the water can be made to fall. The *flow rate* is the quantity of water flowing past a point in a given time. Typical flow rate units are litres per second or cubic metres per second. The *head* is the vertical height, in metres, from the turbine up to the point where the water enters the intake pipe or penstock. The potential power can be calculated as follows:

> Theoretical power (P) = Flow rate (Q) x Head (H) x Gravity (g) = 9.81 m/s2).
> When Q is in cubic metres per second, H in metres and g = 9.81 m/s2) then,
> P = 9.81 x Q x H (kW)

However, energy is always lost when it is converted from one form to another. Small water turbines rarely have efficiencies better than 80%. Power will also be lost in the pipe carrying the water to the turbine, due to frictional losses. By careful design, this loss can be reduced to only a small percentage. A rough guide used for small systems of a few kW rating is to take the overall efficiency as approximately 50%. Thus, the theoretical power must be multiplied by 0.50 for a more realistic figure. *A turbine*

generator set operating at a head of 10 metres with flow of 0.3 cubic metres per second
will deliver approximately, (9.81 x 0.5 x 0.3 x 10 =) 18 kilowatts of electricity.

If a machine is operated under conditions other than full-load or full-flow then other
significant inefficiencies must be considered. Part flow and part load characteristics of
the equipment needs to be known to assess the performance under these conditions.

System Components
Small run-of-the-river hydropower systems consist of these basic components:
- Water conveyance-channel, pipeline, or pressurized pipeline (penstock)
  that delivers the water
- Turbine or waterwheel transforms the energy of flowing water into
  rotational energy
- Alternator or generator transforms the rotational energy into electricity
- Regulator controls the generator
- Wiring delivers the electricity.

Many systems also use an inverter to convert the low-voltage direct current (DC)
electricity produced by the system into 120 or 240 volts of alternating current (AC)
electricity (alternatively you can buy household appliances that run on DC electricity).
Some systems also use batteries to store the electricity generated by the system,
although because hydro resources tend to be more seasonal in nature than wind or solar
resources, batteries may not always be practical for hydropower systems. If you do use
batteries, they should be located as close to the turbine as possible because it is difficult
to transmit low-voltage power over long distances.

Before water enters the turbine or waterwheel, it is first funnelled through a series of
components that control its flow and filter out debris. These components are the
headrace, fore bay, and water conveyance (channel, pipeline, or penstock).

The *headrace* is a waterway running parallel to the water source. A headrace is
sometimes necessary for hydropower systems when insufficient head is provided. They
often are constructed of cement or masonry. The headrace leads to the *fore bay*, which
also is made of concrete or masonry. It functions as a settling pond for large debris that
would otherwise flow into the system and damage the turbine.

Water from the fore bay is fed through the *trash rack*, a grill that removes additional
debris. The filtered water then enters through the controlled gates of the spillway into
the water conveyance, which funnels water directly to the turbine or waterwheel. These
channels, pipelines, or penstocks can be constructed from plastic pipe, cement, steel and
even wood. They often are held in place aboveground by support piers and anchors.

Dams or diversion structures are rarely used in micro hydro projects. They are an added
expense and require professional assistance from a civil engineer. In addition, dams
increase the potential for environmental and maintenance problems.

Figure 104: Hydro electric stations at Russian rivers.

5.12.2 Turbines and water wheels

The water wheel is the oldest hydropower system component. Water wheels are still available, but they are not very practical for generating electricity because of their slow speed and bulky structure.

Turbines are more commonly used today to power small hydropower systems. The moving water strikes the turbine blades, much like a water wheel, to spin a shaft. But turbines are more compact in relation to their energy output than water wheels. They also have fewer gears and require less material for construction. There are two general classes of turbines: impulse and reaction.

Figure 105: Three concepts of hydropower converters: Traditional steel water wheel in southern Germany (left); 2 kW micro turbine (middle); 800 kW Francis turbine from Syzran, Russia (right).

Impulse

Impulse turbines, which have the least complex design, are most commonly used for high head micro hydro systems. They rely on the velocity of water to move the turbine wheel, which is called the runner. The most common types of impulse turbines include the Pelton wheel and the Turgo wheel.

The Pelton wheel uses the concept of jet force to create energy. Water is funnelled into a pressurized pipeline with a narrow nozzle at one end. The water sprays out of the nozzle in a jet, striking the double-cupped buckets attached to the wheel. The impact of the jet spray on the curved buckets creates a force that rotates the wheel at high efficiency rates of 70 to 90 percent. Pelton wheel turbines are available in various sizes and operate best under low-flow and high-head conditions.

The Turgo impulse wheel is an upgraded version of the Pelton. It uses the same jet spray concept, but the Turgo jet, which is half the size of the Pelton, is angled so that the spray hits three buckets at once. As a result, the Turgo wheel moves twice as fast. It's also less bulky, needs few or no gears, and has a good reputation for trouble-free operations. The Turgo can operate under low-flow conditions but requires a medium or high head.

Another turbine option is called the Jack Rabbit (sometimes referred to as the Aquair UW Submersible Hydro Generator). The Jack Rabbit is the drop-in-the-creek turbine, mentioned earlier, that can generate power from a stream with as little as 13 inches of water and no head. Output from the Jack Rabbit is a maximum of 100 W, so daily output averages 1.5 to 2.4 kilowatt-hours, depending on your site.

Reaction
Reaction turbines, which are highly efficient, depend on pressure rather than velocity to produce energy. All blades of the reaction turbine maintain constant contact with the water. These turbines are often used in large-scale hydropower sites.

Because of their complexity and high cost, they aren't usually used for micro hydro projects. An exception is the propeller turbine, which comes in many different designs and works much like a boat's propeller.

Propeller turbines have three to six usually fixed blades set at different angles aligned on the runner. The bulb, tubular, and Kaplan tubular are variations of the propeller turbine. The Kaplan turbine, which is a highly adaptable propeller system, can be used for micro hydro sites.

5.12.3 Pumps as Substitutes for Turbines

Conventional pumps can be used as substitutes for hydraulic turbines. When the action of a pump is reversed, it operates like a turbine. Since pumps are mass-produced, you will find them more readily less expensive than turbines.

However, for adequate pump performance, your micro hydro site must have fairly constant head and flow. Pumps are also less efficient and more prone to damage.

Depending on the end use requirements of the generated power, the output from the turbine shaft can be used directly as mechanical power or the turbine can be connected to an electrical generator to produce electricity. For many rural industrial applications shaft power is suitable (for food processing such as milling or oil extraction, sawmill, carpentry workshop, small scale mining equipment, etc.), but many applications require conversion to electrical power.

For domestic applications electricity is preferred. This can be provided either:

- directly to the home via a small electrical distribution system or,
- can be supplied by means of batteries which are returned periodically to the power house for recharging – this system is common where the cost of direct electrification is prohibitive due to scattered housing (and hence an expensive distribution system)

Figure 106: Control room (left); generator hall with four units in Russian hydropower station (right).

Where a generator is used alternating current (AC) electricity is normally produced. Single-phase power is satisfactory on small installations up to 20 kW, but beyond this, 3-phase power is used to reduce transmission losses and to be suitable for larger electric motors. An AC power supply must be maintained at a constant 50 or 60 cycles/second for the reliable operation of any electrical equipment using the supply. This frequency is determined by the speed of the turbine that must be very accurately governed.

The best geographical areas for exploiting small-scale hydro power are those where there are steep rivers flowing all year round, for example, the hill areas of countries with high year-round rainfall, or the great mountain ranges and their foothills, like the Andes and the Himalayas. Islands with moist marine climates, such as the Caribbean Islands, the Philippines and Indonesia are also suitable. Low-head turbines have been developed for small-scale exploitation of rivers where there is a small head but sufficient flow to provide adequate power.

To assess the suitability of a potential site, the hydrology of the site needs to be known and a site survey carried out, to determine actual flow and head data. Hydrological information can be obtained from the meteorology or irrigation department usually run by the national government. This data gives a good overall picture of annual rain patterns and likely fluctuations in precipitation and, therefore, flow patterns. The site survey gives more detailed information of the site conditions to allow power calculation to be done and design work to begin. Flow data should be gathered over a period of at least one full year where possible, so as to ascertain the fluctuation in river flow over the various seasons. There are many methods for carrying out flow and head measurements and these can be found in the relevant texts.

5.12.4   Economic Issues

Hydro systems generally have a long project life. Equipment such as turbines can last 20 – 30 years, while concrete civil works can last 100 years. This is often not reflected in the economic analysis of small hydropower projects, where costs are usually calculated over a shorter period of time. This is important for hydro projects, as their initial capital costs tend to be comparatively high because of the need for civil engineering works.

Although significant potential exists for further small-scale hydropower, SSH, development, the availability of suitable new sites is limited, particularly if dams or other structures must be built and where local land use and planning laws may limit such development.

A substantial number of weirs and other in-stream structures are in place already and can be retrofitted with hydro equipment. About 3,000 MW of these low-cost applications are estimated to exist globally. The additional environmental and land use impacts of these projects are often very low.

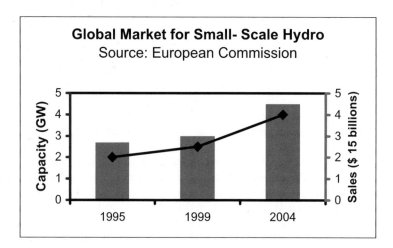

Figure 107: Global Market for Small Scale-Hydro

Hydro developers generally need to invest in detailed analyses before a project can proceed. Regulatory authorities may require structures or systems that prevent adverse effects on flora and fauna, particularly fish. Conversely, some hydro systems may enhance local environments through, for example, the creation of wetlands.

5.12.5   Industry and Market Trends

Over one hundred companies manufacture SSH systems with the most active industries being those in Europe and China. Already China has an installed capacity of 20,000 megawatts and is planning to install 1,500-2,000 MW per year in the period 2001- 2005. Southeast Asia and Latin America are also promising markets.

Although SSH is a mature technology, there is considerable scope for improvement. Electronic controls, telemetry-based remote monitoring, new plastic and anti-corrosive materials, new variable speed turbines for use in low-head applications, and new ways to minimize impacts on fauna, particularly fish, are helping to make systems more cost-effective and extend the range of potential sites.

5.13 Geothermal Energy

Geothermal energy is the energy contained in the heated rock and fluid that fills the fractures and pores within the earth's crust. It originates from radioactive decay deep within the Earth and can exist as hot water, steam, or hot dry rocks.

Geothermal energy is the natural heat of the earth. Enormous amounts of thermal energy are continuously generated by the decay of radioactive isotopes of underground rocks and stored in our globe's interior. This heat is as inexhaustible and renewable as solar energy. The temperature in the core of the earth is in the order of 4,000°C, while active volcanoes erupt lava at about 1,200°C and thermal springs, numerous on land and present on the oceanic floor, can reach 350°C. Presently geothermal energy is exploited by producing the underground water stored in permeable rocks from which it has absorbed available heat (hydro-thermal systems) or, in certain types of geothermal heat pumps, extracting heat directly from the ground. In another approach, still in the experimental stage, hot rocks are artificially fractured and water is let in and circulated between injection and producer wells, gathering, on the way, the rock heat (HDR systems). Still further away is exploitation of the large quantities of heat stored locally at accessible depth in molten rock (magma). Up to 100°C underground water can provide at present, energy for many applications, ranging from district heating to individual residential heating, to agricultural and spa uses and for selected industries. Geothermal fluids between 100° and 150°C can (besides direct heat uses) generate electricity with special (binary) power plants. Above 150°C, the optimal use of the resource is for electricity production.

Hydrothermal resources are renewable within the limits of equilibrium between off take of reservoir water and natural or artificial recharge. It has been calculated (Megel, Rybach 1999) that the life of a low-temperature system (exploited by a couple of producing-injector wells) can extend over more than 150 years, provided alternating periods of production and are adopted. In the commercial development of high-temperature fields, however, the resource is produced for economic reasons at a level exceeding the recharge rate, thus exhausting the fluids, while leaving much heat under the ground.

While production costs are at times competitive and in other cases marginally higher than conventional energy, front-end investment is quite heavy and not easily funded.

5.13.1 Technology

Exploration
Geological, geochemical, and geophysical techniques are used to locate geothermal resources.

Drilling
Drilling for geothermal resources has been adapted from the oil industry. Improved drill bits, slim hole drilling, advanced instruments, and other drilling technologies are under development.

Direct use
Geothermal hot water near the Earth's surface can be used directly for heating buildings and as a heat supply for a variety of commercial and industrial uses. Geothermal direct use is particularly favoured for greenhouses and aquaculture.

Geothermal Heat Pumps
Geothermal heat pumps, or ground-source heat pumps, use the relatively constant temperature of soil or surface water as a heat source and sink for a heat pump, which provides heating and cooling for buildings.

Advanced Technologies
Advanced technologies will help manage geothermal resources for maximum power production, improve plant-operating efficiencies, and develop new resources such as hot dry rock, geopressured brines, and magma.

Electricity generation
Underground reservoirs of hot water or steam, heated by an up welling of magma, can be tapped for electrical power production. To generate electricity, fluids above 150°C are extracted from underground reservoirs (consisting of porous or fractured rocks at depths between a few hundred and 3,000 metres) and brought to the surface through production wells. Some reservoirs yield steam directly, while the majority produces water from which steam is separated and fed to a turbine engine connected to a generator. Some steam plants include an additional flashing stage. The used steam is cooled and condensed back into water, which is added to the water from the separator for reinjection (Figure 108). The size of steam plant units ranges from 0.1 to 150 MW_e.

If the geothermal resource has a temperature between 100° and 150°C, electricity can still be generated using binary plant technology. The produced fluid heats, through a heat exchanger, a secondary working fluid (isobutane, isopentane or ammonia), which vaporises at a lower temperature than water. The working fluid vapour turns the turbine and is condensed before being reheated by the geothermal water, allowing it to be vaporised and used again in a closed-loop circuit (Figure 109).

The size of binary units range from 0.1 to 40 MW_e. Commercially, however, small sizes (up to 3 MW_e) prevail, often used modularly, reaching a total of several tens of MW_e installed in a single location. The spent geothermal fluid of all types of power plants is generally injected back into the edge of the reservoir for disposal and to help maintain pressure.

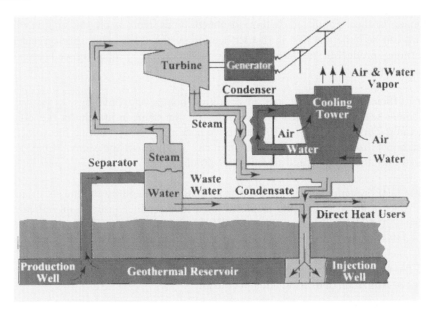

Figure 108: Flash Steam Power Plant (Source: Geothermal Energy, 1998, University of Utah)

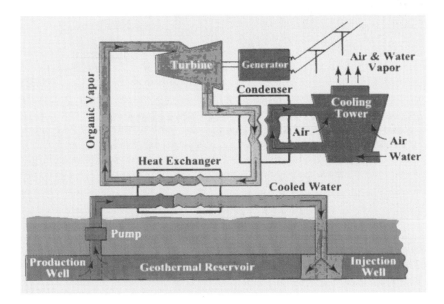

Figure 109: Binary Cycle Power Plant (Source: Geothermal Energy, 1998, University of Utah)

In the case of direct heat utilisation, the geothermal water produced from wells (which generally do not exceed 2,000 metres) is fed to a heat exchanger before being reinjected into the ground by wells, or discharged at the surface. Water heated in the heat exchanger is then circulated within insulated pipes that reach the end-users. The network can be quite sizeable in district heating systems. For other uses (greenhouses, fish farming, product drying and industrial applications) the producing wells are next to the plants serviced.

A very efficient way to heat and air-condition homes and buildings is the use of a geothermal heat pump (GHP) that operates on the same principle as the domestic refrigerator. The GHP (Figure 110) can move heat in two ways: during the winter, heat is withdrawn from the earth and fed into the building; in the summertime, heat is removed from the building and stored under-ground.

In some GHP systems heat is removed from shallow ground by the means of an anti-freeze/water solution circulating in plastic pipe loops (either inserted in vertical wells less than 200 m deep which are then backfilled or buried horizontally in the ground). In other GHP systems flow water produced from a shallow borehole through the heat pump, discharges the water either in another well or at surface. The heat pump unit sits inside the building and is coupled either with a low-temperature floor or wall heating radiator or with a fan delivering hot and cold air.

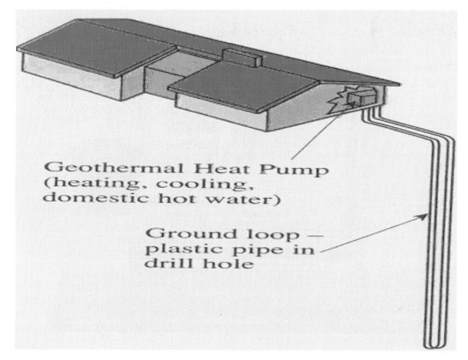

Figure 110: Geothermal Heat Pump (Source: Geothermal Energy, 1998, University of Utah)

5.13.2 Location of Resources

Worldwide, those hot areas with fluids above 200°C at economic depths for electricity production are concentrated in the young regional belts. They are the seats of strong tectonic activity, separating the large crystal blocks in which the earth is geologically divided (Figure 111). The movement of these blocks is the cause of mountain building and trench formation. The main geothermal areas of this type are located in New Zealand, Japan, Indonesia, Philippines, the western coastal Americas, the central and eastern parts of the Mediterranean, Iceland, the Azores and eastern Africa.

Elsewhere in the world, underground temperatures are lower but geothermal resources, generally suitable for direct-use applications, are more widespread. Exploitable heat occurs in a variety of geological situations. It is practically always available in the very shallow underground where GHPs can be installed.

The risk for a prospector (of not locating hot water in the quantity and with the quality required) is limited in shallow depth targets where prior knowledge gained from earlier surveys is available. There are greater uncertainties on deeper resources where insufficient survey work has been conducted.

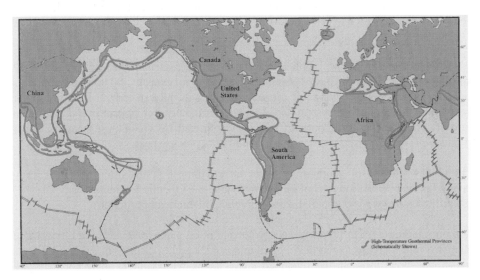

Figure 111: World High Temperature Geothermal Provinces
(Source: Geothermal Energy, 1998, University of Utah)

5.13.3 Recent Developments

Comparing statistical data for end-1996 (SER 1998) and the present survey, it can be seen that there has been an increase in world geothermal power plant capacity (+9%) and utilisation (+23%) while direct heat systems show a 56% additional capacity, coupled with a somewhat lower rate of increase in their use (+32%).

Geothermal power generation growth is continuing, but at a lower pace than in the previous decade, while direct heat uses show a strong increase compared to the past.

Going into some detail, the six countries with the largest electric power capacity are: USA with 2,228 MW_{el} is first, followed by Philippines (1,863 MW_{el}); four countries (Mexico, Italy, Indonesia, and Japan) had ultimo 1999 capacities in the range of 550-750 MW_{el} each. These six countries represent 86% of the world capacity and about the same percentage of the world output, amounting to around 45,000 GWh_{el}.

The strong decline in the USA in recent years, due to overexploitation of the giant Geysers' steam field, has been partly compensated by important additions to capacity in several countries: Indonesia, Philippines, Italy, New Zealand, Iceland, Mexico, Costa Rica, and El Salvador. Newcomers in the electric power sector are Ethiopia (1998), Guatemala (1998) and Austria (2001). In total, 22 nations are generating geothermal electricity, in amounts sufficient to supply 15 million houses.

Concerning direct heat uses, the three countries with the largest amount of installed power: USA (5,366 MW_t), China (2,814 MW_t) and Iceland (1,469 MW_t) cover 58% of the world capacity, which has reached 16,649 MW_t, enough to provide heat for over 3 million houses. Out of about 60 countries with direct heat plants, besides the three above-mentioned nations, Turkey, several European countries, Canada, Japan and New Zealand have sizeable capacity.

With regard to direct use applications, a large increase in the number of GHP installations for space heating (presently estimated to exceed 500,000) has put this category in first place in terms of global capacity and third in terms of output. Other geothermal space heating systems are second in capacity but first in output. Third in capacity (but second in output) are spa uses followed by greenhouse heating. Other applications include fish farm heating and industrial process heat.

The total world use of geothermal power is providing a contribution both to energy saving (around 26 million tons of oil per year) and to CO_2 emission reduction (80 million tons/year if compared with equivalent oil-fuelled production).

The outstanding growth in world energy consumption since 1996 is caused by a more than two-fold increase in North America and a 45% addition in Asia. Europe also has substantial direct uses but has remained fairly stable; reductions in some countries are being compensated by increases in others.

5.13.4 Future Development

The short to medium term future of geothermal energy is encouraging, providing some hurdles that have recently slowed its growth are overcome. Among them: the Far Eastern economic crisis (especially in Indonesia and Philippines, which had ambitious development plans); the strong production decline at The Geysers field in USA; the extended period of low energy prices. Where possible, actions are being taken to improve the situation. At The Geysers an effluent pipeline (to be completed by 2002) is under construction from the town of Santa Rosa, so as to inject into the reservoir as much wastewater as is being produced, thus increasing the field potential.

Energy prices have increased significantly since the second half of 1999. Plans already drafted at the end of the 1990's, but partly delayed, by Indonesia, Philippines and Mexico aim at an additional 2,000 MW_{el} before 2010. In the direct use sector, China has the most ambitious target: substitution of 13 million tons of polluting coal with geothermal energy.

Improved use of hydrothermal resources, limitation of front-end costs and increased ground heat extraction are the keys to a steady development of conventional geothermal energy. Installation of a large number of binary power plants will increase electricity production from wide geographical areas underlain by medium-temperature resources: a good example is the Altheim plant just inaugurated in Austria, which has added power production to district heating with 106°C water. Heat readily available in spas can be optimised by adding compatible uses. New horizons for geothermal energy can be opened up with fresh applications, for example drinking water production on islands and in coastal areas with scarce resources (e.g. the project starting in 2001 on Milos, Greece). Finally, GHP systems can be replicated in many parts of the world.

The long-range future of geothermal energy depends on HDR systems becoming a technological and economic reality. It has been estimated that the heat resources located at economically accessible depths could support, in North America and Europe, an amount of power generation capacity by HDR systems of the same order or greater than present nuclear capacity.

5.13.5 Environmental Impact

Exacting geothermal energy can have adverse environmental impacts, particularly air pollution from radon gas, hydrogen sulphide, methane, ammonia, and carbon dioxide emissions. Generally, the carbon dioxide emissions of a geothermal power plant are only five percent of the emissions from equivalent fossil fuel power plants. Using geothermal resources can also create substantial thermal pollution from waste heat.

Many of these impacts can be controlled with technology that re-injects waste gases or fluids back into the geothermal well. The area of land impacted by a geothermal development is relatively small and such developments can usually co-exist successfully with other land uses.

Other drawbacks include the problem of mineral deposits on the components, and the need to drill new wells after a few years of use. It is important to note that geothermal energy is renewable only if the rate of extraction is less than the recharge rate. Currently, few geothermal projects for generating electricity meet this requirement.

5.13.6 Market Trends

Geothermal energy has been harvested commercially since the early part of the 20th century. The use of geothermal energy has increased rapidly since 1970 and now occurs in more than 45 countries.

About 9,000 MW of electricity is currently generated from geothermal resources, and the equivalent of 9,000 MW is recovered for direct heating applications. There is significant potential to expand the use of geothermal energy for both electricity generation and industrial heating applications.

Total investment in geothermal energy from 1973 to 1995 was about EUR 22 billion, and the industry continues to grow at about 16 percent per annum in electricity generation and about 6 percent in direct uses.

Currently, Costa Rica, El Salvador, Kenya, and Nicaragua generate 10 to 20 percent of their electricity from geothermal resources, while the Philippines generate 22 percent and plants to add 580 MW in the period of 1999-2008. If present trends continue, geothermal electricity generating capacity could increase from about 10,000 MW at the start of 2000, to 58,000 MW in 2020.

The market for ground-source heat pumps is also growing rapidly. In USA, 300,000 domestic and commercial systems are in operation, and under a current incentive scheme, sales could reach 400,000 annually by 2005.

There is considerable interest and research in technology to generate electricity from hot dry rock geothermal resources. In this process, water is injected into the geothermal well and then recovered as steam, which is used to drive turbine generator sets. As the technology, however, is still experimental it is not yet ready for commercial deployment.

5.14 Marine Energy (General)

5.14.1 Resources

The global marine current energy resource is mostly driven by the tides and to a lesser extent by thermal and density effects. The tides cause water to flow inwards twice each day (flood tide) and seawards twice each day (ebb tide) with a period of approximately 12 hours and 24 minutes (a semi-diurnal tide), or once both inwards and seawards in approximately 24 hours and 48 minutes (a diurnal tide). In most locations the tides are a combination of the semi-diurnal and diurnal effects, with the tide being named after the most dominant type.

The strength of the currents varies, depending on the proximity of the moon and sun relative to Earth. The magnitude of the tide-generating force is about 68% moon and 32% sun due to their respective masses and distance from Earth (Open University, 1989). Where the semi-diurnal tide is dominant, the largest marine currents occur at new moon and full moon (spring tides) and the lowest at the first and third quarters of the moon (neap tides). With diurnal tides, the current strength varies with the declination of the moon (position of the moon relative to the equator). The largest currents occur at the extreme declination of the moon and lowest currents at zero declination.

Further differences occur due to changes between the distances of the moon and sun from Earth, their relative positions with reference to Earth and varying angles of declination. These occur with a periodicity of two weeks, one month, one year or longer, and are entirely predictable (Bernstein et al, 1997).

Generally the marine current resource follows a sinusoidal curve with the largest currents generated during the mid-tide. The ebb tide often has slightly larger currents than the flood tide. At the turn of the tide (slack tide), the marine currents stop and change direction by approximately 180°.

The strength of the marine currents generated by the tide varies, depending on the position of a site on the earth, the shape of the coastline and the bathymetry (shape of the sea bed). Along straight coastlines and in the middle of deep oceans, the tidal range and marine currents are typically low. Generally, but not always, the strength of the currents is directly related to the tidal height of the location. However, in land-locked seas such as the Mediterranean, where the tidal range is small, some sizeable marine currents exist.

There are some locations where the water flows continuously in one direction only, and the strength is largely independent of the moon's phase. These currents are dependent on large thermal movements and run generally from the equator to cooler areas. The most obvious example is the Gulf Stream, which moves approximately 80 million cubic metres of water per second (Gorlov, 1997).

Another example is the Strait of Gibraltar where in the upper layer, a constant flow of water passes into the Mediterranean basin from the Atlantic (and a constant outflow in the lower layer). Areas that typically experience high marine current flows are in narrow straits, between islands and around headlands. Entrances to lochs, bays and large harbours often also have high marine current flows (EECA, 1996). Generally the resource is largest where the water depth is relatively shallow and a good tidal range exists. In particular, large marine current flows exist where there is a significant phase difference between the tides that flow on either side of large islands.

There are many sites worldwide with velocities of 5 knots (2.5 m/s) and greater. Countries with an exceptionally high resource include the UK (E&PDC, 1993), Ireland, Italy, the Philippines, Japan and parts of the United States. Few studies have been carried out to determine the total global marine current resource, although it is estimated to exceed 450 GW (Blue Energy, 2000).

5.14.2 Technologies

Useful energy can be generated from marine currents using completely submerged turbines comprising of rotor blades and a generator. Water turbines work on the same principle as wind turbines by using the kinetic energy of moving fluid and transferring it into useful rotational and electrical energy. The velocities of the currents are lower than those of the wind, however owing to the higher density of water (835 times that of air) water turbines are smaller than their wind counterparts for the same installed capacity.

Marine current energy is at an early stage of development, with only a small number of prototypes and demonstration units having been tested to date. There are no commercial grid-connected turbines currently operating. A number of configurations have been tested on a small scale that is essentially marinised wind turbines. Generally speaking, turbines are either horizontal axis or vertical axis turbines. Variants of these two types have been investigated, including turbines using concentrators or shrouds, and tidal fences.

Horizontal axis turbines (*axial flow turbine*)
This is similar in concept to the widespread horizontal axis wind turbine. Prototype turbines of up to 10 kW have been built and tested using this concept. There are currently plans to install a demonstration machine of 300 kW off the south coast of the United Kingdom (MCT, 2000). Concentrators (or shrouds) may be used around the blades to increase the flow and power output from the turbine. This concept has been tested on a small scale in a number of countries, including New Zealand (Rudkin, 2001).

Vertical axis turbines (*cross flow turbine*)
Both drag and lift turbines have been investigated, although the lift devices offer more potential. The best-known example is the Darrieus turbine with three or four thin blades of aerofoil cross-section. Some stand-alone prototypes have been tested, including a 5 kW Darrieus turbine in the Kurushima Straits, Japan. The concept of installing a number of vertical axis turbines in a tidal fence is being pursued in Canada, with plans to install a 30 MW demonstration system in the Philippines (Blue Energy, 2000).

In order for marine current energy to be utilised, a number of potential problems will need to be addressed, including:

Avoidance of cavitation by reducing tip speeds to approximately 8 m/s. This suggests a turbine with a higher solidity than a wind turbine; Prevention of marine growth building up on the blades or ingress of debris; Proven reliability, as operation and maintenance costs are potentially high; and corrosion resistance, bearing systems and sealing.

Turbines may be suspended from a floating structure or fixed to the seabed. In large areas with high currents, it will be possible to install water turbines in groups or clusters to make up a marine current farm, with a predicted density of up to 37 turbines per square km. This is to avoid wake-interaction effects between the turbines and to allow for access by maintenance vessels (DTI, 1999).

As there are currently no commercial turbines in operation, it is difficult to assess the cost of energy and competitiveness with other energy sources. Initial studies suggest that for economic exploitation, velocities of at least 2 m/s (4 knots) will be required, although it is possible to generate energy from velocities as low as 1 m/s. As the technology matures and with economies of scale, it is likely that the costs will reduce substantially.

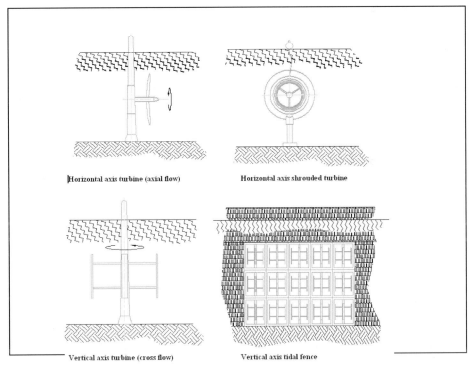

Figure 112: Horizontal and vertical axis turbines

5.14.3   Future of Marine Current Energy

Compared with other renewable technologies, there has been little research into utilising marine current energy for power generation. However, in principle marine current energy is technically straightforward and may be exploited using systems based on proven engineering components (FMP, 1999).

In particular, knowledge gained from the oil and gas industry, the existing hydro industry and the innovative wind energy industry can be used to overcome many of the hurdles facing marine current energy. The global marine current energy resource is very large, and it has a number of advantages over other renewables.

5.15   Ocean Thermal Energy

A great amount of thermal energy (heat) is stored in the world's oceans. Each day, the oceans absorb enough heat from the sun to equal the thermal energy contained in 40 billion tonnes of oil. Ocean Thermal Energy Conversion (OTEC) is a technology, which converts this thermal energy into electricity - often while producing desalinated water.

5.15.1   OTEC Systems

Closed-cycle plants circulate a working fluid in a closed system, heating it with warm seawater, flashing it to vapour, routing the vapour through a turbine, and then condensing it with cold seawater.

Open-cycle plants flash the warm seawater to steam and route the steam through a turbine.
Hybrid plants flash the warm seawater to steam and use that steam to vaporize a working fluid in a closed system.

OTEC systems are also envisioned as being either land-based (or "in-shore"), near-shore (mounted on the ocean shelf), or off-shore (floating) OTEC is a means of converting the temperature difference between surface water of the oceans into useful energy in tropical and sub-tropical areas, and water which comes from the polar regions at a depth of approximately 1,000 metres. For OTEC a temperature difference of 20°C is adequate, which embraces very large ocean areas, and favours islands and many developing countries. The continuing increase in demand from this sector of the world provides a major potential market. Depending on the location of their cold and warm water supplies, OTEC plants can be land-based, floating, or – as a longer term development – grazing. Floating plants have the advantage that the cold water pipe is shorter, reaching directly down to the cold resource, but the power generated has to be brought ashore, and moorings are likely to be in water depths of, typically, 2,000 metres.

The development of High Voltage DC transmission offers substantial advantage to floating OTEC, and the increasing depths for offshore oil and gas production over the last decade suggests that mooring is no longer the problem which it once was – but still a significant cost item for floating OTEC. Land-based plants have the advantage of no power transmission cable to shore, and no mooring costs. However, the cold water pipe has to cross the surf zone and then follow the seabed until the depth reaches approximately 1,000 metres – resulting in a much longer pipe which has therefore greater friction losses, and greater warming of the cold water before it reaches the heat exchanger, both resulting in lower efficiency.

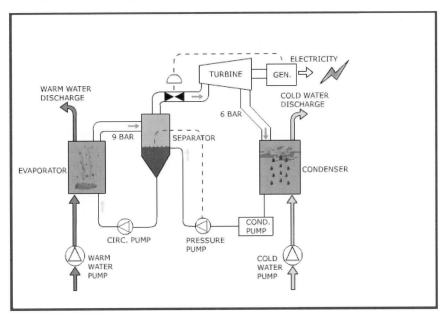

Figure 113: Ocean Thermal Energy Conversion (Source: Petroleum Corporation of Jamaica)

The working cycle may be closed or open, the choice depending on circumstances. All these variants clearly develop their power in the tropical and sub-tropical zones, but a longer-term development – a grazing plant – allows OTEC energy use in highly developed economies which lie in the world's temperate zones. In this case the OTEC plant is free to drift in ocean areas with a high temperature difference, the power being used to split seawater into liquid hydrogen and liquid oxygen. The hydrogen and the oxygen too, in some cases where it is economic, is offloaded to shuttle tankers which take the product to energy-hungry countries. So, in time, the entire world can benefit from OTEC, not just tropical and sub-tropical areas.

A further benefit of OTEC is that, unlike most renewable energies, it is base-load – the thermal resource of the ocean ensures the power source is available day or night, and with only modest variation from summer to winter. It is environmentally benign, and some floating OTEC plants would actually result in net CO_2 absorption. A unique feature of OTEC is the additional products that can readily be provided – food (aquaculture and agriculture); potable water; air conditioning; etc. (see Figure 113). In large part these arise from the pathogen-free, nutrient-rich, deep cold water.

OTEC is therefore the basis for a whole family of Deep Ocean Water Applications (DOWA), which can also benefit the cost of generated electricity. Potable water production alone can reduce electrical generating costs by up to one third, and is itself in very considerable demand in most areas where OTEC can operate.

5.15.2 Environmental and Economic Impact

Calculations for generating costs now take increasing account of "downstream factors" – for example the costs associated with CO_2 emissions. With such criteria included, OTEC/DOWA is becoming an increasingly attractive option. Even without this aspect, the technological improvements (such as the much smaller heat exchangers now required) have contributed to significantly reduced capital expenditure.

On top of these two factors the worldwide trend to whole-life costing benefits all renewables when compared with those energy systems that rely on conventional fuels (and their associated costs), even when the higher initial maintenance costs of early OTEC/DOWA plants are taken into account. When compared with traditional fuels the economic position of OTEC/DOWA is now rapidly approaching equality.

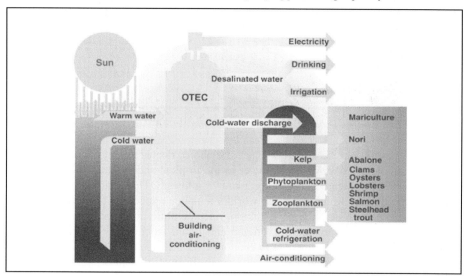

Figure 114: OTEC Applications (Source: US National Renewable Energy Laboratory)

Before OTEC/DOWA can be realised, this R&D must be completed to show clearly to potential investors, via a demonstration-scale plant, that the integrated system operates effectively, efficiently, economically, and safely.

Until such a representative-scale demonstration plant is built and successfully operated, conventional capital funds are unlikely to be available. Whilst the establishment of renewable energy subsidiaries of energy companies is important, there is no doubt that the principal hurdle remaining for OTEC/DOWA is not economic or technical, but the convincing of funding agencies.

In Europe, both the European Commission and the industrially based Maritime Industries Forum examined OTEC opportunities, and in 1997 the UK published its Foresight document for the marine sector, looking five to twenty years ahead. It is significant that the emphasis in the recommendations from all three European groupings has, again, been on the funding and construction of a demonstration. It is recommended to be in the 5-10 MW range and remains the highest single priority.

Figure 115: 210kW OC-OTEC Experimental Plant (1993-1998) in Hawaii
(Source: Luis A. Vega)

A further indication of the interest in DOWA, rather than OTEC alone, is provided by Japan where the industrial OTEC Association was succeeded by the Japan Association of Deep Ocean Water Applications. More recently there has been joint Indian/Japanese work.

The island opportunities have already been mentioned, and in addition to Japan and Taiwan, the European work has stressed these as the best prospects. It is noteworthy that both Japanese and British evaluations have identified Fijian prime sites, one each on the two largest islands of that group.

The worldwide market for renewables has been estimated for the timescales from 1990 to 2020 and 2050, with three scenarios, and all show significant growth. Within those total renewable figures, opportunities exist for the construction of a significant amount of OTEC capacity, even though OTEC may account for only a small percentage of total global electricity generating capacity for some years.

In short, the key breakthrough now required for OTEC is no longer technological or economic, but the establishment of confidence levels in funding agencies to enable building of a representative-scale demonstration plant. Given that demonstration, the early production plants will be installed predominantly in island locations where conventional fuel is expensive, or not available in sufficient quantity, and where environmental impact is a high priority.

5.16 Tidal Energy

Tides are caused by the gravitational attraction of the moon and the sun acting upon the oceans of the rotating earth. The relative motions of these bodies cause the surface of the oceans to be raised and lowered periodically, according to a number of interacting cycles. These include:

- a half day cycle, due to the rotation of the earth within the gravitational field of the moon
- a 14 day cycle, resulting from the gravitational field of the moon combining with that of the sun to give alternating spring (maximum) and neap (minimum) tides
- a half year cycle, due to the inclination of the moon's orbit to that of the earth, giving rise to maximal in the spring tides in March and September
- other cycles, such as those over 19 years and 1,600 years, arising from further complex gravitational interactions.

The range of a spring tide is commonly about twice that of a neap tide, whereas the longer period cycles impose smaller perturbations. In the open ocean, the maximum amplitude of the tides is about one metre. Tidal amplitudes are increased substantially towards the coast, particularly in estuaries. Shelving of the seabed and funnelling of the water by estuaries mainly cause this. In some cases the tidal range can be further amplified by reflection of the tidal wave by the coastline or resonance. This is a special effect that occurs in long, trumpet-shaped estuaries, when the length of the estuary is close to one quarter of the tidal wave length. These effects combine to give a mean spring tidal range of over 11 m in the Severn Estuary (UK). As a result of these various factors, the tidal range can vary substantially between different points on a coastline.

The amount of energy obtainable from a tidal energy scheme therefore varies with location and time. Output changes as the tide ebbs and floods each day; it can also vary by a factor of about four over a spring-neap cycle. Tidal energy is, however, highly predictable in both amount and timing. The available energy is approximately proportional to the square of the tidal range. Extraction of energy from the tides is considered to be practical only at those sites where the energy is concentrated in the form of large tides and the geography provides suitable sites for tidal plant construction. Such sites are not commonplace but a considerable number have been identified in the UK, France, eastern Canada, the Pacific coast of Russia, Korea, China, Mexico and Chile. Other sites have been identified along the Patagonian coast of Argentina, Western Australia and western India.

Tidal energy can also be exploited directly from marine currents induced by the combined lunar and solar gravitational forces responsible for tides. These forces cause semi-diurnal movement in water in shallow seas, particularly where coastal morphology creates natural constrictions, for example around headlands or between islands.

This phenomenon produces strong currents, or tidal streams, which are prevalent around the British Isles and many other parts of the world where there are similar conditions. These currents are particularly prevalent where there is a time difference in tidal cycles between two sections of coastal sea. The flow is cyclical, increasing in velocity and then decreasing before switching to the opposite direction. The kinetic energy within these currents could be converted to electricity, by placing free standing turbo-generating equipment in offshore areas (see Chapter 14: Marine Current Energy).

5.16.1 Technical Concepts

Most countries that have investigated the potential exploitation of tidal energy have concentrated on the use of barrages to create artificial impoundments that can be used to control the natural tidal flow. Barrage developers in the UK and elsewhere concluded that building a permeable barrage across an estuary minimises the cost of civil structures for the quantity of energy that can be realistically extracted.

Construction of barrages across estuaries with high tidal ranges would be challenging but technically feasible. In shallow water armoured embankment would be used, but in deeper water this method would be impractical and too expensive because of the quantity of material required.

Complete closure of estuaries would be achieved by emplacing a series of prefabricated sections, or caissons, made from concrete or steel that could be floated and then sunk into position. The technique has been used in the Netherlands to close the Schelde Estuary. A large steel caisson was used in the construction of the Vadalia power station on a tributary of the Mississippi.

Tidal barrages would comprise sluice gates and turbine generators. Large-scale structures like the Severn Barrages would also include blank caissons and ship-locks. During the ebb tide water is allowed to flow through the sluices and the turbine draft tubes to ensure the maximum possible passage of water into the impounded basin. At or close to high water the sluice gates are closed.

At this stage of the cycle the turbines can be used in reverse as pumps to increase the amount of water within the basin. Although there is an obvious energy demand, the amount of water transferred can provide an additional increase in energy output of up to 10% compared with a cycle where no pumping is used. The actual increase in energy output from pumping depends on the estuary and the tidal conditions. Retention of water allows a head of water (i.e. difference in vertical height of water levels) to be created as the flood tide progresses seaward of the barrage. Once a sufficient head has been created, water is allowed to flow back through the turbines to generate electricity. In this respect a tidal energy barrage is no different to a low-head hydroelectric dam. The large volumes of water and the variation in head require the use of double regulation, or Kaplan turbines. These turbines have guide vanes and blades that can be moved by hydraulic motors. This allows turbine operation, and therefore energy conversion efficiency, to be optimised through each generation cycle as the reservoir head drops.

Experience from the UK's tidal energy programme revealed that ebb generation (i.e. only on the ebb tide) maximises the amount of energy that can be produced from this type of barrage system. Two-way generation (on both the flood and ebb tides) is technically possible, however less energy would be produced because the head of water created prior to generation is lower compared with ebb generation cycle. Moreover, Kaplan turbines in a horizontal configuration are optimised for generation with flow in one direction.

As with all other civil engineering and power generation projects, diligent technical appraisal is essential to mitigate against both technical and commercial risk. Barrage design requires a detailed geotechnical site investigation to determine the foundation conditions. The nature of the substrate and the dimensions of an estuary ultimately determined the design options for barrages. Once an optimal design has been identified, it needs to be developed in detail to establish the construction schedule and the costs at each stage of the project to determine both economic and financial viability. A detailed knowledge of the hydraulic flow pattern before and after the barrage has been constructed is of equal importance and for the same reason.

Hydraulic flow has to be accurately modelled, using complex mathematical models that can accurately simulate natural flow conditions, so that the effects of progressive closure and environmental changes can be predicted. Hydraulic modelling is also used to determine the energy output from the system during each tidal cycle. Other concepts based on secondary artificial storage systems have been investigated, and continue to be promoted. The concept enables storage within two or more basins that can increase the control of the water movement and allows the turbines to operate longer than in single basin schemes. Secondary reservoirs were proposed for the Severn scheme but were discounted because of the cost of the energy produced. The rise in cost is the direct consequence of the substantial additional civil structures required.

5.16.2 Experiences from Current Systems

Tide mills were commonplace along the coasts of Western Europe from the Middle Ages, until the Industrial Revolution supplanted renewable forms of energy with fossil fuel alternatives. Interest in tidal energy was stimulated by the construction of the French barrage across the Rance estuary in Brittany during the 1960's. A dam was built in-situ between two cofferdams. Consequently the entrapped estuarial waters stagnated, although the ecosystem recovered once the barrage began operation. Most of the structure, which has an installed capacity of 240 MW, is comprised of Kaplan turbines with a small bank of sluices. The barrage has a ship rock adjacent to the control centre and carries a trunk road. Originally designed for two-way generation, the operators, EDF, predominantly generate on ebb tides. Despite over thirty years of successful operation, EDF has no plans to build other barrage schemes. Shortly after the completion of the Rance barrage, the Russians built a small experimental system with an installed capacity of 400 kW. The scheme was constructed at Kislogubsk near Murmansk, partly to demonstrate the use of caissons in barrage construction.

The potential for tidal energy at the head of the Bay of Fundy, which extends between the Canadian maritime provinces of Nova Scotia and New Brunswick, has long been recognised. In 1984 a 20 MW plant was commissioned at Annapolis, across a small inlet on the Bay of Fundy's east coast. The barrage was built to demonstrate a large diameter rim-generator (Straflo) turbine. Despite the large tidal energy potential, Canada has relied upon the development of its substantial conventional hydropower reserves. Of more recent interest is Western Australia's tidal energy potential that has been actively promoted near the town of Derby, (situated at the head of two adjacent inlets off the King Sound). The inlets would be connected via an artificial channel. By damming each inlet, differences in water levels in each basin could be controlled which would enable flow via the connecting channel. Power take-off would be achieved from a bank of turbines housed in a structure built in this channel.

Interest in multiple basins has been re-activated in the last three or four years by an American company, Tidal Electric. They are promoting the concept in a number of regions with high tidal ranges including Alaska, Chile and the UK. Water would be moved between three bundled reservoirs built on intertidal mud flats thereby enabling continuous generation. None of these schemes has so far progressed to construction.

5.16.3 Economic Dimensions

Tidal energy projects based on barrages are capital-intensive with relatively high unit costs per installed kilowatt (>EUR 2000/kW). The long construction period for the larger schemes and low load factors would result in high unit costs of energy. The economic performance of tidal energy barrages reflects the influence of site-specific conditions and the necessity for ship locks where access for navigation is required. As barrage construction is based upon conventional technology and site-specific conditions, it is unlikely that significant cost reductions could be achieved. Predicted unit costs of generation are therefore unlikely to change and currently remain uncompetitive with conventional fossil-fuel alternatives.

Some non-energy benefits would stem from the development of tidal energy schemes. However, they would yield a relatively minor monetary value in proportion to the total cost. These benefits are difficult to quantify accurately and may not necessarily accrue to the barrage developer. Employment opportunities would be substantial at the height of construction, with the creation of some permanent long-term employment from associated regional economic development.

5.16.4 Environmental Aspects

Tidal energy barrages would modify existing estuarine ecosystems to varying degrees. Some pre-barrage intertidal areas would become permanently inundated and although the intertidal zonation would change it would still be present and capable of supporting an estuarine ecosystem. The post-barrage upstream intertidal range would be approximately halved but the effect would progressively diminish upstream of the barrage.

Changes to the hydraulic regime will invariably change patterns of sedimentation, eventually leading to a shift in sediment (particle size) distribution. There would be some sediment accumulation upstream of the barrage. The amount will depend on the position of the barrage. In estuaries like the Severn, with high sediment loads, this is an important consideration. For this reason the proposed downstream alignment offers an advantage because it would be less vulnerable to sediment accumulation.

Estuaries are of key importance to migratory species of fish, many of which are the foundation for commercial fisheries. Barrages could act as barriers to migration and damage fish. There is no clear indication from studies on existing hydroelectric stations of the numbers of fish that might be affected. The changes to fish populations are uncertain: levels may fall by 30-50% before the effects of a barrage become evident. Generic R&D has focused on the suitability of acoustic deterrence methods, which will require further refinement.

Much of the site-specific and generic R&D in the UK has concentrated on ornithological studies of migratory birds that use British estuaries in large numbers. Studies have confirmed that bird populations fluctuate between years and within a single winter. Their distribution is also highly uneven, which is partly due to the highly variable distribution of invertebrates. Post-barrage survival rates will depend on the extent of suitable intertidal areas and climatic conditions.

5.17 Wave Energy

The total power of waves breaking on the world's coastlines is estimated at 2 to 3 million MW. In favourable locations, wave energy density can average 40 MW per kilometre of coastline.
Ocean mechanical energy is quite different from ocean thermal energy. Even though the sun affects all ocean activity, tides are driven primarily by the gravitational pull of the moon, and waves are driven primarily by the winds. A *barrage* (dam) is typically used to convert tidal energy into electricity by forcing the water through turbines, activating a generator.

For wave energy conversion, there are three basic systems: channel systems that funnel the waves into reservoirs, float systems that drive hydraulic pumps, and oscillating water column systems that use the waves to compress air within a container. The mechanical power created from these systems either directly activates a generator or transfers to a working fluid, water, or air, which then drives a turbine/generator.

Figure 116: Waves breaking and washing up at the sand beaches of the North Sea (left); the Wave Dragon test module of 20 kW is based on the wash-up principle. In 2003 during testing in Nissum Bredning at the Folkecenter Test Station (right).

5.17.1 Resource

The highest energy waves are concentrated off the western coasts in the 40°–60° latitude range north and south. The power in the wave fronts varies in these areas between 30 and 70 kW/m with peaks to 100kW/m in the Atlantic SW of Ireland, the Southern Ocean and off Cape Horn.

The capability to supply electricity from this resource is such that, if harnessed appropriately, 10% of the current level of world supply could be provided. Work is still needed to determine how much more may be captured by other products (such as pumped water for desalination or electrolysis), once the storage technology for hydrogen is suitably developed.

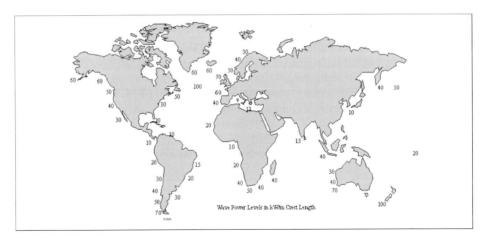

Figure 117: Global Distribution of Deep Water Wave Power Resources

5.17.2   Technologies

Once again, the technologies outlined in 1998 based on Oscillating or Assisted Water
Columns (OWC), buoys and pontoons (the Hosepump), flaps and tapered channels (the
Pendulor and TAPCHAN) still exist or continue to be developed.

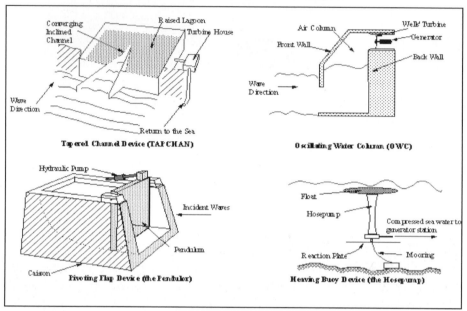

Figure 118: Outline of main types of currently deployed wave energy devices.

Some new developments include:

- The Pelamis (named after a sea-snake), is a series of cylindrical segments
  connected by hinged joints. As waves run down the length of the device and actuate
  the joints, hydraulic cylinders incorporated in the joints to pump oil to drive a
  hydraulic motor via an energy-smoothing system. Electricity generated in each joint
  is transmitted to shore by a common sub-sea cable. The slack-moored device will
  be around 130m long and 3.5m in diameter. The Pelamis is intended for general
  deployment offshore and is designed to use technology already available in the
  offshore industry. The full-scale version has a continuously rated power output of
  0.75 MW.

- Energetech of Australia has developed a two-way turbine that is claimed to be
  significantly more efficient than the Wells turbine. This will be utilised in an OWC
  device that employs a parabolic funnel to focus the wave fronts into the shoreline
  device for greater power capture.

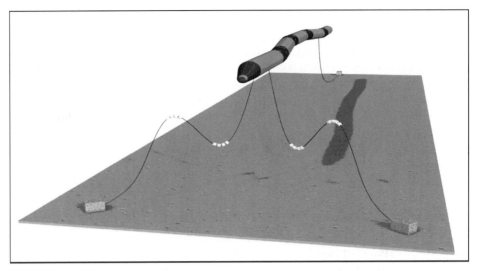

Figure 119: The Pelamis Wave Energy Converter (Ocean Power Delivery Ltd.)

Figure 120: Pelamis – prototype (Ocean Power Delivery Ltd.)

- Denmark has several concepts with very diverse and innovative elements that have been subject to research and development within a governmental programme. The programme was initiated in 1997 and suspended by change of government in 2002. A committee of experts guided more than 40 wave machine inventions through the three phases of the programme, providing equal opportunities to all concepts and inventors. During a short time span a vigorous wave energy sector emerged with sharing of experiences, coordinated by the Danish Wave Energy Association.   In the initial phase all developers could, in principle, obtain a grant of EUR 7,000 for building and testing their models. The most successful concepts obtained in phase two received up to EUR 70,000 and professional R&D standards were required. A few projects proceeded to phase three and obtained grants of up to EUR 1.5 million before the programme was closed by change of government. Therefore a number of promising concepts did not get the opportunity to demonstrate their development potential. Two of the earliest projects were:

A. The Waveplane – which is a wedge-shaped structure that channels incoming waves into a spiral trough thus producing a vortex to drive a turbine. A one-fifth-scale model without a turbine has been tested in a fjord in Jutland in mid-1999.
B. The Wave Dragon – which is a floating tapchan using a pair of curved reflectors to gather waves over the top to a ramped trough where water is released though a low-head turbine. A one-fifth-scale model has been tested; a quarter-scale prototype was deployed in April 2003 for testing at the Wave Energy Test Station operated by the Folkecenter for Renewable Energy in North-western Denmark. The full-size device (estimated to have a generation peak of 4 MW) is large, with a "span" across the reflector arms of 227m.

Figure 121: Folkecenter Test Station for Wave Energy Machines in Nissum Bredning (left): Wave Rotor by Eric Rossen and Ecofys (middle); Swinging Cylinder by Öjvind Boltz (right).

•   In the USA, a company called Ocean Power Technologies (OPT) based in New Jersey is utilising a sheet of piezo-electric polymer material which, when deflected mechanically, produces electricity directly.

The technology scene for wave power is becoming more vibrant as various techniques and devices continue to be developed. It is evident that the range of types of devices are far from exhausted, thus providing encouragement for the future.

5.17.3 Economy of Wave Power

Projections for the cost per unit of electricity for various wave devices were made in 1998. They show off-shore and near-shore devices producing power in the EUR 0.07

– 0.10 /kWh range (based on 8% discount rate). The trends shown in the same report show a halving in the predicted cost over a period of six or seven years. This is borne out by the experience of onshore wind energy costs, which have been seen to fall by a factor of five over 12 to 15 years. Based on these results, it is reasonable to expect that wave energy unit costs can be made to fall to the EUR 0.03 – 0.04/kWh range within 3 to 5 years.

Following these trends will depend on several factors including:
- the ability of developers, manufacturers and installers to engineer-out cost from devices, especially as greater numbers are manufactured and deployed in arrays;
- the commitment of governments and local authorities to streamline planning and regulatory processes;
- the development of suitable approaches to grid connection, both for smaller "embedded" supplies and major power sources. This requires governments, electricity distributors and the financial community to collaborate in new ways;
- the flow of innovation from R&D on more cost-effective materials, design and construction methods;
- mechanisms being made available (under national electricity regulation regimes) to support specific emerging technologies with access to long-term contracts and/or to include wave power in capital grant mechanisms while the technologies mature;
- the ability of the wave power industry to show good practice in standardised independent testing and performance assessment methods from an early stage;
- the willingness of the financial community to recognise the key role of renewable energy technologies (including wave energy conversion) as a significant future proportion of the energy balance and to seek positively to invest into it.

5.18 Hydrogen

Hydrogen is the third most abundant element on the earth's surface, where it is found primarily in water (H_2O) and organic compounds. It is generally produced from hydrocarbons or water; and when burned as a fuel, or converted to electricity, it joins with oxygen to again form water. Hydrogen is the simplest element; an atom consists of only one proton and one electron. It is also the most plentiful element in the universe.

Despite its simplicity and abundance, hydrogen does not occur naturally as a gas on the Earth - it is always combined with other elements. Water, for example, is a combination of hydrogen and oxygen (H_2O). Hydrogen is also found in many organic compounds, notably the "hydrocarbons" that make up many of our fuels, such as gasoline, natural gas, methanol, and propane.

Hydrogen is produced from sources such as natural gas, coal, gasoline, methanol, or biomass through the application of heat; from bacteria or algae through photosynthesis; or by using electricity or sunlight to split water into hydrogen and oxygen.

Hydrogen can be made by separating it from hydrocarbons by applying heat, a process known as "reforming" hydrogen. Currently, most hydrogen is made this way from natural gas. An electrical current can also be used to separate water into its components of oxygen and hydrogen. Some algae and bacteria, using sunlight as their energy source, give off hydrogen under certain conditions.

Fuel cells are a promising technology for use as a source of heat and electricity for buildings, and as an electrical power source for electric vehicles. Although these applications would ideally run off pure hydrogen, in the near term they are likely to be fuelled with natural gas, methanol, or even gasoline. Reforming these fuels to create hydrogen will allow the use of much of our current energy infrastructure - gas stations, natural gas pipelines, etc. - while fuel cells are phased in.

In the future, hydrogen could also join electricity as an important *energy carrier*. An energy carrier stores, moves, and delivers energy in a usable form to consumers. Renewable energy sources, like the sun, cannot produce energy all the time. The sun does not always shine. But hydrogen can store this energy until it is needed and can be transported to where it is needed.

Some experts think that hydrogen will form the basic energy infrastructure that will power future societies, replacing today's natural gas, oil, coal, and electricity infrastructures. They see a new *hydrogen economy* to replace our current energy economies.

Figure 122: Ford Focus with 2 litre motor converted to wind-hydrogen by Folkecenter (left); sustainable living: Hydrogen car, windmill and straw bale house (right).

5.18.1 Technologies

Technologies with the best potential for producing hydrogen to meet future energy demand fall into four general categories:

Thermochemical

A steam reforming process is currently used to produce hydrogen from such fuels as natural gas, coal, methanol, or even gasoline. To draw on renewable energy sources, the gasification or pyrolysis of biomass - organic material - can be used to generate a fuel gas that can be reformed into hydrogen.

Electrochemical

The electrolysis of water produces hydrogen by passing an electrical current from renewable resources through it. Here wind energy is an extremely promising source of supply of the necessary electricity.

Photo electrochemical

The photo electrochemical (PEC) process produces hydrogen in one step, splitting water by illuminating a water-immersed semiconductor with sunlight.

Photo biological

Photobiological systems generally use the natural photosynthetic activity of bacteria and green algae to produce hydrogen.

5.18.2 Transport and Storage

The use of hydrogen as a fuel and energy carrier will require an infrastructure for safe and cost-effective hydrogen transport and storage. Future experiments should deal with centralized and decentralized production structures utilizing the best practices within the renewable energy fields.

Figure 123: Hydrogen storage for Ford Focus with 10x9 litres composite bottles of 200 bar (left); Hydrogen storage mounted in luggage room of car (middle); filling station with compressor and 900 litres storage of 225 bar (right).

Hydrogen is currently stored in tanks as a compressed gas or cryogenic liquid. The tanks can be transported by truck or the compressed gas can be sent across distances of less than 200 kilometres by pipeline as is being practiced by the industries in the Ruhr district in Germany.

Technologies that store hydrogen in a solid state are inherently safer and have the potential to be more efficient than gas or liquid storage. These are particularly important for vehicles with on-board storage of hydrogen. Technologies under investigation include:

- Metal hydrides that involve chemically reaction to the hydrogen with a metal.

- Carbon nanotubes taking advantage of the gas-on-solids adsorption of hydrogen.
- Glass microspheres relying on changes in glass permeability due to temperature, to fill the microspheres and trap hydrogen.

5.18.3 Economy

The vision of building an energy infrastructure that uses hydrogen as an energy carrier - a concept called the "hydrogen economy" - is considered the most likely path toward a full commercial application of hydrogen energy technologies.

5.19 Fuel Cells

5.19.1 The Technology

A fuel cell converts chemical energy directly into electricity by combining hydrogen and oxygen in a controlled reaction.

Figure 124: Fuels cell stack by Ballard (left); hydrogen fuel cell 5 kw module for family house demonstrated at Tokyo Gas (middle); Hydrogen fuel stack by GAT (right).

Fuel cells emit virtually no pollution, as the waste "exhaust" is simply water vapour and heat. In many applications, the waste heat can be used, making a fuel cell system much more efficient than conventional power supplies. In combined applications, fuel cell systems can, similar to cogeneration, convert 80 % of the energy available in the fuel into electrical and heat energy, however with an increased electricity share.

William Grove first conceived the idea of a fuel cell in 1839, some 40 years before the invention of the internal combustion engine. There are five basic technologies under development for both stationary and mobile applications by more than 30 major private companies, including major automobile manufacturers.

Although the fuel cell is not a renewable energy technology per se, it can certainly be a core element in a renewable energy system, particularly if the hydrogen comes from a renewable fuel or process, such as a biofuel or electrolysis via solar-generated electricity. Fossil fuels like natural gas can also be reformed for use in fuel cells with relatively low emissions. Gas from coal and diesel fuel is poor choices of fuel since they cause an overall increase in CO_2 emissions. With the hydrogen being supplied from solar energy, wind power or biomass the combined system can be the ultimate power source. By combining hydrogen and oxygen to produce electricity and heat, the "exhaust" from a fuel cell is simply water vapour. Put the reaction in reverse - use electricity from a renewable resource to split water into hydrogen and oxygen - and a complete, cyclic, and a virtually non-polluting process can create both electricity and heat.

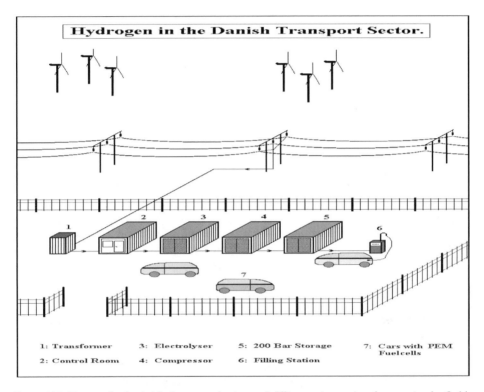

Figure 125: Decentralized wind-hydrogen production and filling station project for cars in the Sydthy municipality will utilise the surplus of wind electricity.

The different technologies use a number of methods to control the re-combination of hydrogen and oxygen, including membranes. Fuel cell types under development include proton exchange membrane (also called 'polymer electrolyte membrane' or PEM), phosphoric acid, alkaline, molten carbonate, and solid oxide fuel cells.

The different fuel cell types can be divided into two groups: low temperature, or first generation (alkaline fuel cell, solid polymer fuel cell and phosphoric acid fuel cell); and high temperature, or second generation (molton carbonate fuel cell and solid oxide fuel cell). Low temperature cells have been commercially demonstrated, but are restricted in their fuel supply and are not readily integrated into combined heat and power applications. The high temperature cells can use a wide variety of fuels though internal reforming techniques, have a high electrical efficiency, and can be integrated into combined heat and power systems.

Methanol (MeOH) and hydrogen (H_2) derived from biomass have the potential to make major contributions to transport fuel requirements by competitively addressing all of these challenges, especially when used in fuel cell vehicles (FCVs).

In a fuel cell, the chemical energy of fuel is converted directly into electricity without first burning the fuel to generate heat or run a heat engine. The fuel cell offers, similar to cogeneration, a leap in energy efficiency. Dramatic technological advances, particularly for the proton exchange membrane (PEM) fuel cell, focus attention on this technology for motor vehicles.

Fuel cells are under development to provide power in applications ranging from a few watts (to power a cell phone, for example) to tens of megawatts (a district power supply). They are inherently modular and can be expanded to suit different applications.

5.19.2 Motive Power

For motive applications, particularly road transport, the PEM fuel cell is the current technology leader used by almost all major vehicle manufacturers. PEM fuel cells have the advantage of operating at low temperatures (about 80° C). Besides using pure hydrogen by using special "reforming" technology, virtually any hydrogen-rich fuel can be used in PEM cells, including methanol, propane, natural gas, and gasoline. However, fuel cells with such reformers are more costly and complex than fuel cells using pure hydrogen.

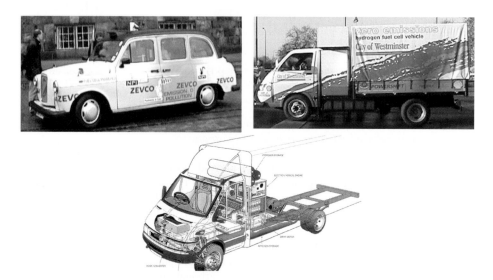

Figure 126: Vehicles with alkaline fuel cells by ZEVCO: London taxi being demonstrated in front of Danish Parliament (left); IVECO delivery van in London (right and bottom).

Ballard Power Systems, of Vancouver, Canada, introduced a prototype PEM fuel cell bus in 1993 and plans to introduce PEM fuel cell buses on a commercial basis beginning in 1998. In April 1994, Germany's Daimler Benz introduced a prototype PEM fuel cell light duty vehicle (a van) and announced plans to develop the technology for commercial automotive applications. A range of prototypes of cars and buses have appeared without, however, becoming commercially available.

Figure 127: Fuel Cell electric bus, Georgetown, USA, by UTC Fuel Cells (left). 2ⁿᵈ generation fuel cell transit buses in Chicago, USA, by Ballard Power Systems (right).

Part of the automobile industry is planning intermediate hydrogen development strategies by converting the robust and cheap piston engine to hydrogen operation. With a price for fuel-cells of EUR 10,000 per kW (2003), the ordinary piston engine may have a future fuelled by hydrogen as if it was operating on petrol, diesel or propane and renewable energy in the form of biogas, ethanol, plant oil etc.

Table 29:   Main types of fuel cell

Fuel cell	Electrolyte	Operating temperature (°C)	Development status	Applications
SOFC	Solid oxide	750-1000	100 kW tubular 5 kW planar	CHP, power generation
PEMFC or SPFC	Solid polymer	50-80	250 kW	Transport, CHP, distributed power generation
AFC	Alkaline	50-200	Developed	Space, transport
MCFC	Molton carbonate	630-650	2 MW	Power generation, CHP
PAFC	Phosphoric acid	190-210	11 MW	CHP, power generation

5.19.3 Stationary Power

A number of firms already offer commercial systems and more firms plan to offer residential fuel cell systems in 2002. These systems will likely use a PEM fuel cell to produce both power and domestic hot water in a casing not much larger than a conventional hot water system. Furthermore, the overall efficiency of this type of system is significantly higher than separate heat and power production and with a higher share of electricity compared to advanced cogeneration units. In these innovative systems, excess electricity may be exported into the local power grid markets, with an increasing demand for the "green" electrons.

For larger commercial applications, phosphoric acid, solid oxide, or molten carbonate fuel cells are suitable with the added benefit that the waste heat (up to 400° C) can be used for cogeneration of heat or cooling raising overall efficiency.

5.20 Stirling Engines for Power and Heat Generation

The idea of the Stirling engine was conceived in the 19th century. The name comes from its implementation of the Stirling cycle. Nitrogen or helium gas (Weber, 1987) is shuttled back and forth between the hot and cold ends of the machine by the displacer piston. The power piston, with attached permanent magnets, oscillates within the linear alternator to generate electricity.

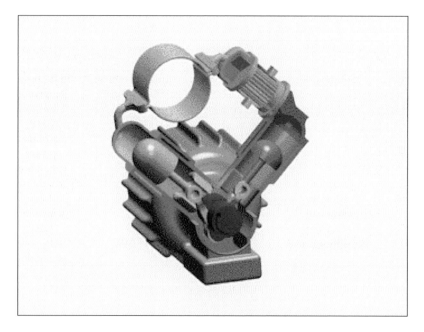

Figure 128: Principal construction of a Stirling engine (SOLO-Stirling, 2003)

New developments by Sunpower Inc. (1997) indicate that their Stirling engine's generator efficiently converts rough biomass into electricity with intrinsic load matching capacity. A single engine produces 2.5 kW of 60Hz, 120V alternating current power, and up to four engines may be used with a single burner. It is proposed for use at domestic and light industrial sites, and may be used for cogeneration, delivering both electricity and heat.

Positive features of the engine include the direct conversion of low-grade biomass and solar heat into electricity, and an integrated generator with a combined heat and power efficiency up to 90%.

5.20.1 The Technology

In contrast to Otto and Diesel engines, the operating chambers of the Stirling engine are sealed. Heat is supplied from an external source. Fossil fuels such as oil or gas can be used as well as re-generative solar energy and biomass. Combustion residue cannot penetrate the interior of the engine with the clear advantages of low wear and long maintenance intervals.

Compared with conventional, larger CHP units, smaller CHP units powered by Stirling engines do have distinct advantages. The sealed engine design of Stirling engines might allow maintenance intervals between 5000 to 8000 operating hours. Operating costs are considerably lower than for gas driven Otto engines (SOLO, 2003).

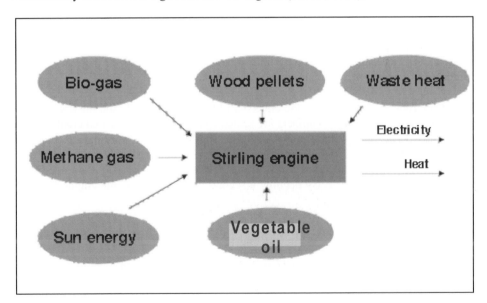

Figure 129: Using Re-generative Energy Sources

The 90° V-2-Cylinder engine is built of a compression and an expansion-cylinder in which the working gas is moved in a closed thermodynamic cycle. Inside the compression-cylinder the gas is isothermally compressed at a low temperature level by cooling with water, and then moved through the regenerator.

Usually Helium is used as working gas due to the good thermal and aerodynamic properties. Due to the closed cycle with heat transfer from the outside through the heater, the Stirling engine is independent from the heat source. If a burner is used, the flue gases leave the combustion chamber with a temperature of approx. 800 °C.

Figure 130: Stirling generator (SOLO, 2003)

To obtain good efficiency, the thermal energy has to be transferred to the combustion air by an air preheater, where the air is heated up to 600 °C. For this reason, burners for efficient Stirling engines differ by the pre-heating of air (from normal heating combustors) working at a much higher temperature level of 1200-2000 °C.

The piston rods are connected to the crankshaft by connecting rods and the dry-running pistons in the high-pressure chambers are sealed against the oil-lubricated crankcase by piston seals. The output performance can be adjusted by the working gas pressure between 40 and 150 bars in the range of 3 to 10 kW (mechanical). This is realized by a small piston-pump which pumps the working gas from the engine to a storage bottle with a higher pressure level. By opening a second magnetic valve, the engine pressure can be adjusted.

5.20.2 Applications

Stirling engines can be used for power as well as for heat generation.

The solar version is based on the design for gas-heated engines. The site at Almeria commenced operation in 1992 and successfully completed more than 40,000 operating hours.

Figure 131: Solar stirling generator (SOLO-Stirling, 2003)

Countries with high solar radiation can harness that energy source. A Stirling engine operates efficiently at the focal point of a parabolic dish. Further development of these types of engines is necessary and will open significant future opportunities to produce electricity directly from biomass.

Table 30 summarizes the power ranges and efficiencies of the biomass conversion technologies mentioned in this chapter.

Table 30: Power range and reported efficiencies of different electricity technologies (Grimm, 1996).

Technology	Power range (MWe)	Typical overall efficiency (%)
Steam engine	0.025-2.0	16
Steam turbine (back pressure)	1–150	25
Steam turbine (extraction-condensation cycle)	5-800	35
Steam turbine (condensation cycle)	1-800	40
Gas Otto motor	0.025-3.5	25-43
Gas turbine	1.0-200	35
Integrated gas and steam combined cycle	5-450	55
Stirling engine	0.0003-?	40
Fuel cell	0.005-?	50
Combined Heat and Power	0.005-200	90

5.21 Tractors as power generator (dual uses)

The En-O-Trak System (Schluchter 2002), accesses the unused potentials of generating added value by means of tractors. The innovative idea lies in creating additional modules that allow tractors to produce heat and electric power. Electricity generation is possible due to the principle of power and heat takeoff. The system can work utilizing both conventional and alternative energy sources, thus achieving high autonomy and independence from the supplier of the electric power and power transmission lines. Electric power generation using alternative energy sources also allows saving taxes on the emission of pollutants into the environment

The introduction of this system will allow tractors to contribute significantly to the return of capital investments by producing heat and electric power. In the ideal case, during scheduled agricultural work, machines can be used for their immediate purposes, and in their spare time - for earning additional profit by generating heat and electric power.

Figure 132: Massey-Ferguson tractor using PPO as fuel is prepared for heat and power generation.

5.21.1 Technological process

The module consists of a heat-tolerant, noise-isolated portable container that is attached to the tractor with such components as a generator, a heat exchanger, a control unit, a power supply system, ventilation, exhaust output system with exhaust purification etc..

In the initial phase En-O-Trak modules are specially designed to work with tractors for the reason that their design is perfectly suited for hooking up with the module. The tractor provides simple connection to all systems of the engine and is an integral part of any agricultural enterprise. Moreover, the tractor provides for high reliability in all operational conditions.

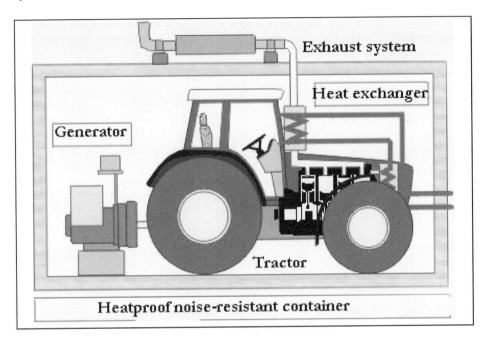

Figure 133: Basic elements of the module for heat and power production.

Kinetic energy take-off from a tractor is enabled by connecting the generator to the front or back shaft of the engine. Energy takeoff requires higher expenses so changes do not affect the tractors performance with respect to its primary goal (" a normal mode "). It is also essential to prevent air or dirt from entering the engine cooling system while equipping the module. The principle of heat extraction remains standard - through the gas exhaust system or the cooling system radiator. Besides, the heated exhaust of the module can be used for heating hangars, garages etc.

The current system based on a tractor with a 100 KW engine, operating at a 90% load, generates 70 KW of electric energy and 89 KW of heat. Per hour this system uses either 22 litres of vegetable oil or 31 m³ of biogas plus 2 litres of compression ignition fuel per hour that may also be renewable energy. Thus, 33 % of the energy input is converted into electricity and 42 % into useable heat. Currently, 25 % of the energy input is lost as unused heat. In the course of technical improvements the efficiency of electric power generation should rise.

5.21.2 Fuel

En-O-Trak Modules can use the energy sources shown in Table 31 as fuel. It is possible to use various kinds of fuel. The power supply system of a tractor is adapted to the use of different kinds of fuel.

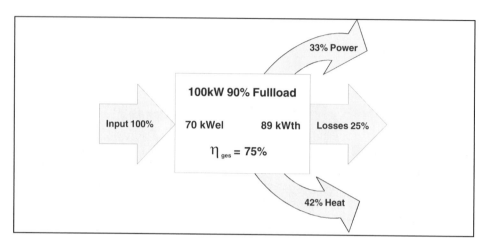

Figure 134: Energy flow diagram.

Table 31: Kinds of fuels for the operation of the tractors
www.enotrak.de

Liquid	Gaseous
Vegetable oil	Biogas
Biodiesel	Gas from waste products
Gastronomic oil	Gas from sewer waste products
Black oil	Wood gas
Diesel fuel	Natural gas (fossil)

Such energy sources as vegetable oil, biodiesel, biogas, and wood gas can be produced at the farm.

Using standard and alternative energy sources, optimum use of agricultural land, increased tractor utilization, we find reduced CO_2 emissions with an accelerated return on the tractor investment. With the use of vegetable oil as fuel we find compensation for 68 % of the cost of the tractor. If biogas is used, this will help not only provide 100 % return on investment, but also some additional profit.

5.22 Batteries

5.22.1 Types of Batteries

Since electric light is normally needed on demand, the only way of ensuring this is to provide electrical storage in the form of a battery. It is possible to use primary batteries (i.e. dry cells) that are charged when bought and thrown away when exhausted. These are convenient but extremely expensive in terms of electrical energy costs and are particularly expensive in the tropics since transport and distribution costs are incurred in moving them to the rural areas. Their application is limited to torches, electronic equipment, communication, toys etc.

Secondary batteries, which are rechargeable, are more cost-effective than primary batteries, but an investment in equipment to recharge them becomes necessary. The two main secondary battery options available are lead-acid and nickel-cadmium batteries; the former being similar to the batteries used in cars. Nickel-cadmium batteries are generally less available (except as dry-cell substitutes) and cost more, but they can be more robust and tolerant of abuse than lead-acid batteries. However they self-discharge quite quickly if not used. Electrical energy from a lead-acid battery can cost as little as one twentieth to one fifth as much for the same amount of energy delivered from primary (dry) batteries.

Figure 135: Two truck batteries for PV-supply of village school in West Africa (right); 6 Volt heavy-duty batteries by Surrette in home power installation (left).

For most lighting purposes lead-acid batteries are probably the easiest and cheapest secondary cell option. They are available as deep-discharge batteries that have a longer life than car batteries, and if looked after tend to be better for general electrical storage. Most lead-acid and nickel-cadmium batteries require regular checking of their electrolyte level and topping off with distilled or deionised water, (not with acid). Rainwater can be used for this purpose, providing it has not been contaminated. Low maintenance and maintenance-free lead-acid batteries are also available, at slightly increased cost.

An important point to note with lead-acid batteries is that their life is considerably shortened if they are over-discharged. Ideally they should only be discharged to about

50% of their full rating; i.e. a 60Ah (ampere-hour) battery should only be discharged to 30Ah before recharging it.

Batteries are generally provided with nominal voltages in multiples of 2V; common larger capacity lead-acid batteries will be 12V or 24V nominal voltage. Filament bulbs can be obtained which operate at almost any battery voltage, but fluorescent and vapour discharge lamps need 'mains voltage' (typically 230 Volt AC) This can be readily provided by interfacing an appropriate inverter between the battery and the lamp; the high efficiency of such lamps make it well worth investing in an inverter if long hours of light are needed. Some small fluorescent lights in fact include a built-in inverter, and these often run at higher frequency than mains frequency that improves the light's efficiency still further. In most cases it will be best to obtain a fluorescent light with a built-in inverter so it can be directly connected to a 12V or 24V dc (battery) source.

5.22.2 Sources for Charging Batteries

There are four main methods by which the electricity for charging may be provided:

- taking the battery to the nearest mains supply and putting it on charge
- using a small engine-powered generating set
- using a photovoltaic charging system (solar cells)
- using a small wind generator

Using a petrol generating set imposes considerable problems as, unless power is being generated for other purposes too, the charging current, which is acceptable for small battery storage for just one or two lights is rather low for even the smallest generating sets; hence the engine needs to be run at part load which results in inefficient fuel use and is bad for the engine.

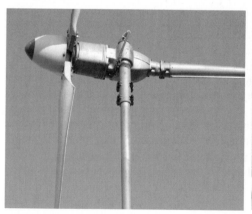 

Figure 136: Mass produced 500 watt Chinese windmill for battery charging (left); a variety of crystalline and thin film PV panels can be used for battery charging (right).

A photovoltaic system is by far the most universally applicable, as adequate sun for charging can be found in most parts of the world, and such a system, apart from occasional cleaning of the array, requires almost zero attention. The snag is that photovoltaic arrays currently cost (delivered and installed) in the region of £3-5/Wp (peak Watt).

The supplier should be able to advice on the size needed (and can generally supply a battery and lighting system too). To run a 30W light for six hours would typically require, in a sunny tropical location, two nominal 40W solar modules. Areas with extended cloudy periods may need up to twice this capacity. The last option is to use a small wind generator; this will generally be a cheaper option than solar power in locations having mean wind speeds greater than 4.0m/s in the least windy months.

5.23 Alternative Transportation Fuels

Alternative fuels are being used today in place of gasoline and diesel fuel made from petroleum. The following fuels as "alternative transportation fuels": biodiesel, electricity, ethanol, hydrogen, methanol, natural gas, propane, p-series, and solar energy. Using these alternative fuels can help nations reduce dependence on imported petroleum and improve air quality.

Figure 137: 3-wheel bicycle with electric motor support.

Pure Vegetable Oil (PPO, PVO)
Pure vegetable oils (palm oil, rape oil or sunflower oil) are being used as fuels for specially modified automotive engines.

Biodiesel
Biodiesel is a diesel alternative that can be made from vegetable oils, animal fats, and even recycled cooking greases. Biodiesed is classified similar to diesel in terms of soil

protection and health.The health effects of biodiesel are low, its emissions are low, and it's biodegradable. Biodiesel is the product of esterification of bio-oils.

Electricity

Electricity is considered a fuel when used in electric vehicles. Electricity as a fuel shifts the burden of pollution control to the electrical supply systems, resulting in much lower automobile emissions per kilometer travelled.

Figure 138: Small electric cars charged from wind power: By Think (left); by Kewett (right).

Ethanol

Mainly used today as a fuel additive (see Improved Petroleum-Based Fuels), ethanol is also used in an 85-percent-ethanol/15-percent-gasoline blend, called E85. The main technical goals are to lower the cost of ethanol while expanding the ethanol infrastructure. Currently, the industry is supported by various fuel standards, codes, and legislation.

Hydrogen

Although hydrogen can fuel an engine directly, or serve as a fuel additive, the current emphasis is on the use of hydrogen to supply fuel cells, which power electric vehicles. Hydrogen has also been blended with methane to form a fuel called Hythane.

Methanol

Like ethanol, methanol is blended with gasoline in a ratio of 85 to 15, to form M85. When burned in an engine, methanol produces low emissions. Methanol is also the fuel for the direct-methanol fuel cell. Developing a methanol-refuelling infrastructure is essential to expand the use of M85.

Natural Gas

Natural gas (fossil gas) is an alternative to conventional fuels. It is used in vehicles as compressed natural gas (CNG) or liquefied natural gas (LNG). At issue is the availability of refuelling sites for vehicles that run on these fuels.

Propane

Propane is usually used in the form of liquefied petroleum gas (LPG). Again, the availability of refuelling sites is an issue for vehicles that run on this fuel.

P-series

P-series fuels are new fuels that are now classified as an alternative fuel.

P-series fuels are blends of ethanol, methyltetrahydrofuran (MTHF), natural gas liquids and butane. The ethanol and MTHF are expected to be derived from renewable domestic feedstocks, such as corn, waste paper, cellulosic biomass, agricultural waste and wood waste from construction. The P-series fuels emissions standards. P-series fuels join the list of alternatives to gasoline that includes ethanol (E85), methanol (M85), natural gas, propane and electricity.

Innovative regenerative transportation fuels can be produced from various sources:

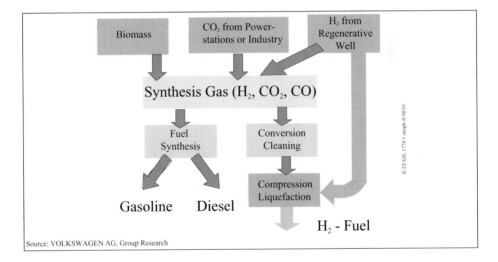

Figure 139: Regenerative Fuels

Table 32: *Properties of various fuels*
www.nerl.gov

Properties of fuels

Property	Gasoline	No. 2 Diesel Fuel	Methanol	Ethanol	MTBE	Hydrogen
Chemical Formula	C_4 to C_{12}	C_3 to C_{25}	CH_3OH	C_2H_5OH	$[CH_3]_3COCH_3$	H_2
Composition, Weight-%						
Carbon	85 to 88	84 to 87	37.5	52.2	66.1	0
Hydrogen	12 to 15	33 to 16	12.6	13.1	13.7	100
Oxygen	0	0	49.9	34.7	18.2	0
Density, kg/m^3 @ 15.56°C	719 to 779	803 to 887	794	792	742	-
Boiling temperature, °C	26.7 to 225	187.8 to 443.3	65	77.8	55	-253
Reid vapour pressure, bar	0.55 to 1.03	0.01	0.32	0.16	0.54	-
Research octane no.	90 to 100	-	107	108	116	130
Motor octane no.	81 to 90	-	92	92	101	-
Freezing point, °C	-40	-40 to -34.4	-92.5	-114	-108.9	-259.4
Viscosity, Pa*s @ 15.56 °C	0.00037 to 0.00044	0.0026 to 0.0041	0.00059	0.00119	0.00035	-
Flash point, closed cup, °C	-42.8	73.9	11.1	12.8	-25.6	-
Autoignition temperature, °C	257.2	appr. 316	463.9	422.8	435	-17.2
Flammability limits, volume-%						
Lower	1.4	1	7.3	4.3	1.6	4.1
Higher	7.6	6	36	19	8.4	74
Latent heat of vapourization KWh/m^3 @ 15.56 °C	appr. 70	appr. 54	258.61	184.13	66.82	-

Property	Gasoline	No. 2 Diesel Fuel	Methanol	Ethanol	MTBE	Hydrogen
Chemical Formula	C_4 to C_{12}	C_3 to C_{25}	CH_3OH	C_2H_5OH	$[CH_3]_3COCH_3$	H_2
Heating value Higher (liquid fuel- liquid water), KWh/m³	9760.9	10739.4	4974.8	6279.5	-	-
Lower (liquid fuel- water vapour), KWh/m³ @ 15.56 °C	8904.34	9941.9	4398	5884.61	7239.62	3
Higher (liquid fuel- liquid water), KWh/l	9.67	10.74	4.97	6.28	-	-
Lower (liquid fuel- water vapour), KWh/l @ 15.56 °C	8.9	9.94	4.4	5.88	7.24	-
Higher (liquid fuel- liquid water), KWh/kg	12.21 to 13.18	12.41 to 12.92	6.3	8.27	11.82	39.42
Lower (liquid fuel- water vapour), KWh/kg @ 15.56 °C	11.63 to 12.28	11.63 to 12.28	5.54	7.43	9.76	33.3
Heating value, stoichiometric mixture Mixture, vapour state, KWh/m³ @ 15.56 °C	0.99	1	0.96	0.96	-	-
Fuel in liquid state, KWh/kg	0.83	-	0.86	0.83	-	-
Specific heat, kJ/kg*K	1.12	1	1.4	1.33	1.16	-
Stoichiometric air/fuel, weight	14.3	14.7	6.45	9	11.7	34.3
Volume-% in vapourized stoichiometric mixture	2	-	12.3	6.5	2.7	-

6 Applications of Renewable Energy Technologies

6.1 Cookers and Stoves

6.1.1 Solar Cookers

In developing countries, a solar cooker can provide basic cooking energy in areas of high solar radiation. In these areas, simple solar stills can also be used to purify water. These devices can be very simple and constructed using local materials and labour.

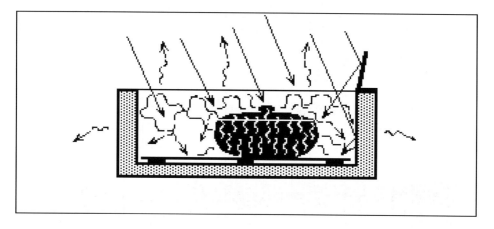

Figure 140: The principals of the function of a box solar cooker.

Single or multiple reflectors bounce additional sunlight through the glass and into the solar box. This additional input of solar energy results in higher cooker temperatures.

Much higher temperatures can be achieved with parabolic cookers. The parabolic dish systems concentrate solar radiation to a single point to produce high temperatures.

In a new development by El Bassam (2001) a solar stove was designed which combines the advantage of the utilization of thermal and electrical power of the solar energy. The materials used are of high-tec quality, which ensure a durable and efficient utilization. Several companies participated in this project.

Figure 141: Different solar cookers in Africa and China (El Bassam 2000)

Figure 142: The combined thermal and electrical solar cooker demonstrated in Alsharja (Alnajar, 2002)

6.1.2   Liquid Stove

Liquid fuels can be used in wick and pressure stoves. Due to their high viscosity, plant oils cannot be used in common wick stoves. Therefore, a pressure-cooking stove has been developed for plant oils (Stumpf and Mühlbauer 2002).

In those stoves pressure is induced in a tank through application of a pump. The liquid evaporates in a vapourizer and is emitted through a nozzle into the combustion area. The jet rebounds at a rebounding plate, mixes with ambient air and burns in a blue flame. The combustion area is surrounded by a flame holder. The power output is adjusted with a valve regulating the fuel flow. For ignition, a small amount of ethanol is incinerated in a pre-heating dish beneath the vapourizer.

Figure 143: Liquid fuel stove

6.1.3   Wood Stoves

The improvement of traditional stoves has been a focus of much of the research and development work carried out on biomass technologies for rural areas of developing countries. This was initially in response to the threat of deforestation but has also been focused on the needs of women to reduce fuel collection times and improve the kitchen environment by smoke removal.

There have been many approaches to stove improvement, some carried out locally and others as part of wider programmes run by international organisations. Figure 143 below shows a variety of successful improved stove types, some small, portable stoves and others designed for permanent fixture in a household.

Some of the features of these improved stoves include:

- a chimney to remove smoke from the kitchen
- an enclosed fire to retain the heat
- careful design of pot holder to maximise the heat transfer from fire to pot
- baffles to create turbulence and hence improve heat transfer
- dampers to control and optimise the air flow
- a ceramic insert to minimise the rate of heat loss
- a grate to allow for a variety of fuel to be used and ash to be removed
- metal casing to give strength and durability
- multi pot systems to maximise heat use and allow several pots to be heated simultaneously

Figure 144: Various types of wood stoves

Improving a stove design is a complex procedure that requires a broad understanding of many issues. Involvement of users in the design process is essential to gain a thorough understanding of the user's needs and requirements for the stove. The stove is not merely an appliance for heating food (as it has become in Western society), but is often acts as a social focus, a means of lighting and space heating. Tar from the fire can help to protect a thatched roof, and the smoke can keep out insects and other pests.

Cooking habits need to be considered, as well as the lifestyle of the users. Light charcoal stoves used for cooking meat and vegetables are of little use to people who have staple diets such as Ugali, which require large pots and vigorous stirring. Fuel type can differ greatly; in some countries cow dung is used as a common fuel source, particularly where wood is scarce. Cost is also a major factor among low-income groups. Failing to identify these key socio-economic issues will ensure that a stove programme will fail. The function of an improved stove is not merely to save fuel.

6.2   Air Conditioning, Cooling, Refrigeration and Ice Making

There is a demand for cooling in many parts of the world where there is no firm electricity supply, and conventional fuels are difficult or expensive to obtain. Requirements tend to be either for medical uses, where a high capital cost per kW of cooling is acceptable, or for food (especially fish) preservation where the cooling power required is much greater and the acceptable cost per kW may be lower.

Vaccine storage refrigerators have been sold at a cost of EUR 90-270/W cooling [1], the lower cost being for a solar thermal system sold by the French company BLM and the larger for a typical photovoltaic system. These high costs are considered acceptable since the application is related to medical provision. Harvey [2] has shown that in the case of a fish storage ice-maker for Zambia, the required capital cost was EUR 30/W and he concluded that a 1 tonne ice/day solar thermal refrigerator could be built to this price. There is little possibility that the higher costs per watt of the smaller units could be justified in larger plants. To achieve cost-effective air conditioning from solar energy is much more difficult. The target for an economic system is about EUR 1/W.

6.2.1   Possible refrigeration cycles

There are six cycle classes that can be used for renewable powered air conditioning systems:

Electrical driven
A standard mechanical vapour compression cycle, requiring an electrical input to a compressor. The electricity can be generated by photovoltaic panels, wind and hydro generators. Solar, thermal or biomass steam generators are other options. This has the advantage of using off-the-shelf technology, but the disadvantages of high cost and the probable need for an electricity storage sub-system.

Intermittent adsorption cycles
Adsorption refrigeration cycles rely on the adsorption of a refrigerant gas into an adsorbent at low pressure and subsequent desorption by heating. The adsorbent acts as a 'chemical compressor' driven by heat. In its simplest form an adsorption refrigerator consists of two linked vessels, one of which contains adsorbent and both of which contain refrigerant.

Intermittent absorption cycles
These are thermodynamically identical to adsorption systems but use liquid absorbents rather than solid adsorbents. Typically the pair used is ammonia-water, but ammonia-NaSCN, methanol-LiBr and other pairs have been used experimentally

Continuous absorption cycles with an electrically driven feed pump eliminates the problems of bulk, but if electricity is available to drive a solution feed pump then it could be argued that it would be better to use a conventional vapour compression cycle.

The use of a small amount of photovoltaic electricity to drive a feed pump might be justified.

The Platen-Munters diffusion absorption cycle is continuous and does not use a mechanical pump. It is used successfully in small gas or kerosene refrigerators and freezers but has proved difficult to adapt to larger sizes and to irregular heat sources such as solar energy.

Desiccant wheel systems
In industrialised countries where there is grid electricity, there is no cost effective application for solar powered refrigeration of food, medicines etc. However, there may still be the possibility of cost-effective solar air conditioning. The target cost per unit of cooling must be competitive with conventional systems, perhaps around EUR1/Watt. One possibility for cost effective solar powered air conditioning is desiccant cooling. A number of solar assisted air conditioning systems using desiccant wheels have been evaluated in Germany [4], Sweden [5] and in the USA where a system has been installed in a Kentucky Fried Chicken Restaurant in Florida [6]. These systems normally use solar energy together with a gas fired heater backup in order to provide a cost-effective solution. The major drawback of a desiccant system in a desert climate is its use of water, which may be an expensive commodity.

Cost targets for solar/biomass powered air conditioning are presented and a number of possible technical solutions are discussed. There are three projects underway at the University of Warwick that could be used for solar thermal air conditioning. They all use active carbon – ammonia adsorption cycles (Critoph, 2003).

The *monolithic intermittent cycle* could be used but with comparatively low efficiency. Its main application is see as an icemaker for food preservation in developing countries.

The *convective thermal wave* system is patented and has been under development for some years. It uses ammonia refrigerant and a granular active carbon adsorbent in a regenerative cycle. The principle has already been proved in a laboratory system, but now the challenge is to build a 10 kW chiller in 1 m³ demonstration unit. The project, funded under a DETR Future Practice award supported by six industrial partners (including a manufacturer of evacuated tube collectors) aims to commission the new machine during 2002. The final target for cooling COP is 0.95 (0.4 from solar) under ARI conditions. The only significant technical risk is in the development of a novel ammonia circulator. There are no other moving parts within the refrigeration system and all other components are well proven conventional technology. The operating results of the previous laboratory system are presented, together with the overall design and performance predictions for the new machine.

The *multiple bed regenerative* cycle is a new concept. It also uses ammonia, but the adsorbent is a monolithic active carbon specially developed for this project. The system is particularly suited to air conditioning and has no moving parts within the refrigeration system itself. The cycle is highly regenerative and the COP could be as high as for the convective wave machine. It is at the research, rather than development, stage with a first laboratory system scheduled for completion in March 2002. The principles of operation and performance predictions are given.

6.2.2   ISAAC™ Solar Icemaker

In February 2002, two ISAAC Solar Icemakers were installed in Upper Egypt for the Ministry of Agriculture and Land Reclamation (MoA) for demonstration to the fishermen. The units were installed at fish hatcheries operated by the MoA, one at Abu Simbel and the other outside of Aswan. The ISAACs installed at Lake Nasser have 16 foot by 8 foot solar collectors. The capacity is about 110 lb of ice per day depending on sunlight and ambient temperature. The ISAAC Solar Icemaker does not require any electricity or fuel and does not use an ozone-depleting refrigerant. The ISAAC uses solar energy to generate refrigerant during the day. During the night refrigerant is absorbed while ice is formed.

Figure 145: Ice formation by ISAAC solar icemaker

By installing ISAACs around Lake Nasser ice can be provided to the fishermen without relying on a distribution system and without using electricity. Ice will increase quality and quantity of fish production and increase income.

A single ISAAC can preserve enough perishables for more than 100 people. At only pennies per day per person, the ISAAC is an economical solution to help feed people in rural areas, eliminate waste in post-harvest handling, preserve resources, and protect the environment (Solar Ice Company, USA 2003).

6.2.3   Active Solar Cooling and Refrigeration

It is possible to use solar thermal energy or solar electricity to operate or power a cooling appliance or a refrigerator. Active solar energy systems use a mechanical or electrical device to transfer solar energy absorbed in a solar collector to another component in the "system". It is possible to also cool a building or structure by using the natural processes of solar heat transfer (conduction, convection, and radiation). This is often referred to as "passive solar cooling", and is primarily an architectural technique. This section focuses on active solar cooling systems.

Absorption Cooling and Refrigeration

Absorption cooling is the first and oldest form of air conditioning and refrigeration. An absorption air conditioner or refrigerator does not use an electric compressor to mechanically pressurize the refrigerant. Instead, the absorption device uses a heat source, such as natural gas or a large solar collector, to evaporate the already-pressurized refrigerant from an absorbent/refrigerant mixture. This takes place in a device called the vapour generator. Although absorption coolers require electricity for pumping the refrigerant, the amount is small compared to that consumed by a compressor in a conventional electric air conditioner or refrigerator. When used with solar thermal energy systems, absorption coolers must be adapted to operate at the normal working temperatures for solar collectors: 82° to 121°C (180° to 250°F). It is also possible to produce ice with a solar powered absorption device, which can be used for cooling or refrigeration.

Desiccant Cooling

Desiccant cooling systems make the air seem cooler by removing most of its moisture. In these systems, the hot, humid outdoor air passes through a rotating, water-absorbing wheel. The wheel absorbs most of the incoming air's moisture. This "desiccates" (heats and dries) the air. The heated air then passes through a rotating heat exchanger wheel, which transfers the heat to the exhaust side of the system. At the same time, the dried air passes through an evaporative cooler, further reducing its temperature. The heated exhaust air continues through an additional heat source (e.g. a solar heat exchanger), raising its temperature to the point that the exhaust air evaporates the moisture collected by the desiccant wheel. The moisture is then discharged outdoors. The various system components require electricity to operate, but they use less than a conventional air conditioner. Most desiccant cooling systems are intended for large applications, such as supermarkets and warehouses. They are also ideal for humid climates.

Evaporative Cooling/Photovoltaic-Powered

Electric evaporative coolers, also known as adiabatic or "swamp" coolers have been common for many years in hot, dry climates. As the outside air passes through a fine mist of water, it gives up much of its heat through evaporation. In *direct* evaporative systems, the evaporation process humidifies the air. In *indirect* evaporative systems, the evaporation process is isolated from the air stream, and uses a heat exchanger to cool the air. It is possible to design a solar photovoltaic (PV) array to provide some or all of the electricity to operate the unit.

Heat Engine/Vapour Compression Cooling (Rankin-Cycle)
The Cycle-cycle cooling process uses a vapour compression cycle similar to that of a conventional air conditioner. Solar collectors heat the working fluid, which has a very low vaporization point. The working fluid then drives a Cycle-cycle heat engine. This technology, however, is mainly experimental, and is not used often because it needs a large system size to do any meaningful amount of cooling.

Photovoltaic (PV)-Powered Heat Pumps, Air Conditioners, and Refrigerators
PV cells/modules can power devices such as evaporative coolers, heat pumps, and refrigerators. In most cases, you need an inverter to change the low-voltage, direct current (DC) produced by the PV array into the higher-voltage, alternating current (AC) that powers most heat pumps, air conditioners, and refrigerators. Only small, off-the-shelf, PV-powered cooling appliances are currently available. It is possible to design a PV array to provide some or all of the electricity to operate these types of systems. Systems used for vaccine refrigeration are becoming widely used in remote areas of developing countries.

6.2.4   Zeolite-Refrigerator

The solar-powered refrigerator (see Figure 146) consists of a cooler box with a built-in evaporator and uses water as the cooling agent. In addition to this, all that is needed is a manually operated vacuum pump and one or more canisters containing the non-toxic mineral zeolite (a naturally occurring mineral) and a parabolic dish (Frank & Zech).

Figure 146: Complete Refrigeration System

To produce the desired cooling effect, one of the zeolite containers is connected up to the evaporator. Air is extracted from the system with the aid of the vacuum pump. The vacuum created by the pump causes the boiling point of the water to drop so rapidly that the water boils at room temperature. The energy needed to bring this about is drawn from the water itself, such that it cools to freezing point. The cooler box will then remain at a temperature from 0 to 6°C for 72 hours. Once the zeolite (which has adsorbed the very high-energy steam) is saturated, the zeolite canister is disconnected from the evaporator and replaced by a canister of non-saturated regenerated zeolite. A few pumps to remove the air from the system once again and the refrigeration process continues. The saturated zeolite is then dried with the aid of solar box or parabolic heater. The mineral can be endlessly recycled in this way.

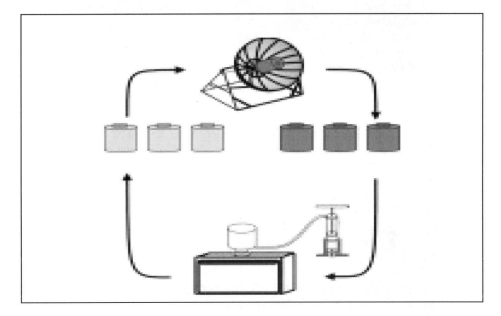

Figure 147: Solar heater for drying saturated zeolite

The advantages of the solar-power refrigerator include: cordless refrigeration (which means that the cooler box can also be used to transport medicines), the use of simple and non-toxic materials, the possibility to produce a refrigerator locally and easily, and for less expense than the conventional type.

6.2.5   *Earthware pots cooler*

The art of pottery is deeply rooted in African culture. Traditionally, all manner of vessels, from cooking pots to wardrobes, which are moulded from clay, have largely been replaced by aluminium containers (Mohammed Bah Abba).

Figure 148: Earth pot for water cooling.

The innovative cooling system that Mohammed Bah Abba developed in 1995 consists of two earthware pots of different diameters, one placed inside the other. The space between the pots is filled with wet sand that is kept constantly moist, thereby keeping both pots damp. Fruit, vegetables and other items such as soft drinks are put in the smaller inner pot, which is covered with a damp cloth and left in a very dry, ventilated place. The water contained in the sand between the two pots evaporates towards the outer surface of the larger pot where the drier outside air is circulating. The evaporation process causes a drop in temperature of several degrees, cooling the inner container and extending the shelf life of the perishable food inside.

Abba carried out several trials with the pot, consistently refining the invention over a two year period. He found that aubergines would stay fresh for 27 days instead of three, tomatoes and peppers lasted for up to three weeks and African spinach, which spoils after one day in the intense tropical heat, remained edible for 12 days.

After several refinements to the prototype pot, Abba was satisfied with the invention and then went about making it available to local rural communities. He employed some of the local unemployed pot makers to produce an initial batch of 5000 pots. These cost about EUR 0.30 per cooler to produce. These pots were distributed, free of change, to five villages in Jigawa, northern Nigeria. Later in 1999, Abba built additional pot-making facilities and produced a further 7,000 pots that were supplied to another 12 local villages. Abba estimates that almost three quarters of rural families in Jigawa are now using his cooling device.

Figure 149: The earthware cooler and the preservation of various vegetables and fruits.

Abba has recently started to sell the pots at EUR 0.40 per pair, which is EUR 0.10 higher than the original production cost. He uses the profit to further develop and expand production. His aim is to export the pot cooler to other hot, dry countries where cold food storage is a problem.

6.3   Solar Powered Village Scale Potable Water System

B.F.Brix.

6.3.1   Introduction

Water and energy are key components of sustainable economic development, and are inextricably linked. Appropriate water treatment methods are essential for developing communities. Systems need a blend of a community based approach with sustainable technologies and education. Despite the fact that water covers sixty percent of the world's surface (of this 97% is salt water and 2% is locked in Polar ice caps), clean, uncontaminated drinking water is becoming our most rare and endangered resource. It has been reported that over 1.5 billion people in the world have no direct access to potable quality water, while four billion of the world's six billion people remain unconnected to collective water sanitation systems. (World Water Forum, 2000). The World Health Organisation (WHO) reports that each year over 5 million people die of water related illnesses including the staggering statistic that every eight seconds a child becomes another casualty. Not only does consumption of poor quality water result in a high health toll, but it has been shown that improving water and sanitation brings valuable benefits to both social and economic development. The WHO has estimated that annually, over ten million person-years of time and effort are provided by women and female children carrying water from distant, often polluted sources.

Water in some regions is limited, while in many regions supplies are adequate, but have become contaminated and therefore harmful for human consumption. There are numerous treatment methods available for purifying contaminated water, but few are suitable for remote area applications particularly in developing countries. This is often due to capital expenditure and ongoing costs, the level of expertise that is needed for plant operation or associated infrastructure that may be necessary. In many remote communities the problem of contaminated water is compounded by the lack of a reliable electricity supply needed to power water treatment technology. Where water supplies have been contaminated by salts and other pollutants, more than 40% of the cost of providing water lies in the energy required to purify it. (Hoffman, 2000)

6.3.2   Implications of Consumption of Contamined Water

Because of the essential role water plays in supporting human life, if contaminated it also has a huge potential to spread a wide variety of illnesses and disease. Water related illnesses are rare in the developed world, mainly due to the fact that water supply and sanitation systems are well managed and financed. In the developing world however, inadequate water supply and sanitation systems result in millions of deaths annually. This results from water that has become contaminated by human, natural or industrial means which can consequently cause a variety of communicable diseases and illness through both ingestion or by physical contact.

6.3.2.1 Water-related disease

Water-borne diseases: include diseases that are spread by the contamination of water by human/ animal faeces or urine. This is the most common form of water related disease and creates the most harm on a global scale, see Table 33. These broad spectrum of diseases are caused by the ingestion of pathogenic bacteria or viruses including Cholera, Giardiasis and Typhoid.

Water-washed diseases: are generally spread by poor personal hygiene. This may be as a result of inadequate water supply for washing or by physical contact with contaminated water. Diseases that fall within this category may not be fatal but still have a serious debilitating effect on sufferers. Diseases of this type include scabies, trachoma, bacterial ulcers and tick-borne diseases.

Water-based diseases: are caused by parasites that depend upon the pathogenic organisms spending part of its life cycle in water. Many diseases that fall within this category are caused by worms and infection often occurs by penetration of the skin rather than by consumption of water. Examples are schistosomiasis (bilharzia) and drancunculiasis (guinea worm).

Water-related insect vectors: diseases are spread by insects that breed or feed near water. Infection of the disease is not connected to consumption or physical contact with the water. Malaria, Yellow Fever, Filariasis and Dengue Fever are examples of diseases that fall within this category and are spread by mosquitoes.

6.3.2.2 Chemical-related illness

In addition to water borne diseases, there is another broad spectrum of illnesses that result from consumption of water that has been contaminated by chemicals. These pollutants usually result from human activities through agriculture and industry but in many cases occur either indirectly or directly via natural causes. These contaminants usually do not pose dire consequences from single exposures but usually produce observable health effects after medium to long-term consumption.

The WHO has published guidelines for drinking water that lists the acceptable concentration of certain contaminants. A number of which pose no direct consequence to health even if ingested at higher levels, but nevertheless may be objectionable to consumers for various reasons. Examples of this include colour, taste and odour that may then lead to a reduced level of intake and consequently other health concerns.

There are many undesirable constituents in groundwater that do cause detrimental health concerns and the WHO guideline levels on these should no be exceeded.

Nitrates
The level of nitrates in many ground waters is a major problem in many areas. Nitrites occur naturally in most ground-waters and many surface waters but concentrations are rising in many areas due to the residual effect of fertilizers used in agriculture.

Consumption of nitrates shows no real heath concern for children or adults, but can be dangerous to bottle fed babies up to the age of six months. If a baby below this age is bottle-fed with milk made up with water containing high levels of nitrates (10-20mg/l) there is a possibility of methaemoglobinaemia (blue baby disease), caused by an oxygen depletion in the blood stream.

Fluoride

Fluoride is added to many urban water supply systems at a level of approximately 1mg/l in an effort to prevent tooth decay. Fluoride occurs naturally in many ground waters at levels exceeding the drinking water guidelines. At levels above 1.5mg/l there is a risk of yellow staining of teeth and at levels higher than this there is a danger of bone damage.

Arsenic

Arsenic occurs naturally in many regions around the world. Its presence is due to the dissolution of minerals and ores, and its concentrations have elevated in some areas to dangerously high levels due to the erosion of local rocks. Long term ingestion of dissolved arsenic in water can lead to cancer of the skin, lungs, bladder and kidneys as well as skin pigmentation.

It can be seen from the above that in order to reduce the toll of water related diseases it is important that improvements be made not only to water supply, but also in efficient water purification/sanitation systems. The WHO reports that improved water and sanitation can reduce morbidity and mortality rates of some of the most serious diseases by as much as 80%. It must also be noted that rather than the installation of sophisticated 'developed-country' style water supply/treatment systems that are often donated through international aid projects, appropriate technology should be used to provide these services in developing regions. Many of these 'un-appropriate' systems are of little long term value unless ongoing operation and maintenance assistance is also provided.

6.3.3   *Appropriate Technology – Water Treatment Methods*

Appropriate technology may be defined as the implementation and dissemination of sustainable technologies and community based approaches that protect natural resources and assist people, especially the economically underdeveloped areas, in becoming more self reliant while also increasing community living standards.

Many suitable technologies exist for the sanitation and purification of a broad range of contaminated water. However many of these are un-suitable for remote area applications particularly in developing countries. This may be due to:
- the level of expertise needed to operate and maintain the systems does not exist
- a lack of infrastructure such as electricity grid
- ongoing cost of specialized consumables may be prohibitive

Following is a brief description of some technologies that have had some success in the remote area applications of water purification.

Table 33: Estimates of Morbidity and Mortality of Water-related Diseases (Source: WHO)

Disease	Morbidity (episodes/year, or as stated)	Mortality (deaths/year)	Relationship of Disease to Water Supply and Sanitation
Diarrhoeal diseases	1,000,000,000	3,300,000	Strongly related to unsanitary excreta disposal, poor personal and domestic hygiene, unsafe drinking water
Infection with intestinal helminths	1,500,000,000$^\alpha$	100,000	Strongly related to unsanitary excreta disposal, poor personal and domestic hygiene
Schistosomiasis	200,000,000$^\alpha$	200,000	Strongly related to unsanitary excreta disposal and absence of nearby sources of safe water
Dracunculiasis	100,000	-	Strongly related to unsafe drinking water
Trachoma	150,000,000$^\beta$	-	Strongly related to lack of face washing, often due to absence of nearby sources of safe water
Malaria	400,000,000	1,500,000	Related to poor water management, water storage, operation of water points and drainage
Dengue Fever	1,750,000	20,000	Related to poor solid wastes management, water storage, operation of water points and drainage
Poliomyelitis	114,000	-	Related to unsanitary excreta disposal, poor personal and domestic hygiene, unsafe drinking water
Trypanosomiasis	275,000	130,000	Related to the absence of nearby sources of safe water
Bancroftian filariasis	72,800,000$^\alpha$	-	Related to poor water management, water storage, operation of water points and drainage
Onchocerciasis	17,700,000$^{\alpha,\chi}$	40,000$^\delta$	Related to poor water management in large-scale projects

Source: WHO Fact sheet No. 112

α People Currently Infected χ Includes an estimated 270,000 blind

β Case of the active disease. Approx. 5,900,000 cases of blindness or severe complications of Trachoma per year.

δ Mortality caused by blindness.

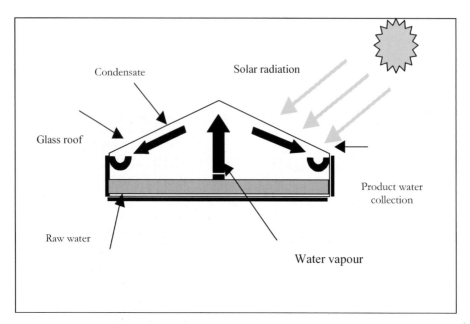

Figure 150:   *A typical solar distillation plant design*

6.3.3.1   Solar Distillation

Solar distillation is a method of water desalination that has been applied in areas of the world that receive relatively high amounts of solar radiation. The main developments in this technology were in the 1960's and 1970's where improvements were made in the efficiency and reductions in construction costs. Over this period many large distillation plants were constructed around the world, however most of them are not in operation today.

Solar distillation is primarily a simple desalination process in which the energy of the sun is used to evaporate water that is contained within a glass covered basin. Water vapor then condenses on the coolest surface, which is normally the stills roof, and is then collected as product water. The salt and other impurities remain in the basin and must be cleaned out on a regular basis. Distillation involves a phase change of the feed solution, so in theory a well-designed system will separate both organic and inorganic contaminants. However this is not the case in most solar distillation systems and an additional disinfection system should be used.

Solar stills can produce approximately 2.5-5 litres/m2 per day, they are simple to operate and maintain. The major drawback of this technology is that it is usually only suitable for small-scale applications because of the large areas of land that must be covered by the relatively fragile, solar receivers. Because the system is essentially modular in concept, it is easy to increase the output of a plant by increasing the number of stills, although the cost increase is fundamentally linear with the quantity of stills.

Hence the cost becomes prohibitive for larger systems, as there is limited economy of scale.

Figure 151: Water-for-Life grey water treatment R&D plant for family houses by Folkecenter.

6.3.3.2   Ultra violet sterilisation

Ultraviolet (UV) sterilisation works by dissociating the DNA structure of living cells thus preventing their multiplication. UV light is a particular selection of wavelengths from the light spectrum. Special lamps may be used to shine a particular wavelength light onto the water to be treated, which results in the killing of bacteria and viruses. The water to be treated must be first filtered to reduce turbidity, this is essential to provide the highest possible light transmission through the water.

The energy requirements for UV sterilisation are relatively low which make it a viable option to power from photovoltaic or other renewable sources. Disadvantages of UV sterilisation are that no residual disinfection is left in the water, which gives no guarantee of post treatment contamination. UV treatment is primarily a method of sterilisation and usually must be used with other water treatment methods to remove other contaminants such as dissolved solids, pesticides etc.

6.3.3.3   Reverse Osmosis

Osmosis is a naturally occurring phenomenon. Simply, if two salt solutions are separated by a membrane, water will pass from the more dilute side to the more concentrated side of the membrane. *Reverse osmosis* (RO) is a process whereby high purity water can be produced by reversing thus natural 'osmotic force' of a solution by pressurizing the water to be treated. High quality water will then permeate though the membrane leaving the salts and other impurities behind.

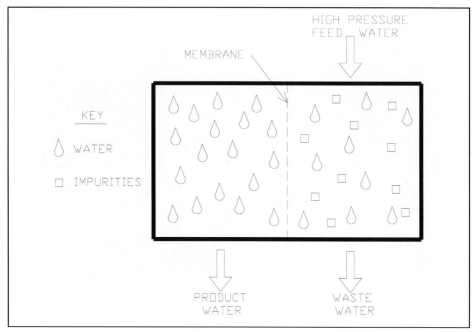

Figure 152: The basic reverse osmosis process

Reverse osmosis is the finest filtration currently available and will allow the removal of particles as small as ions from a solution. Most reverse osmosis technology uses a process known as cross flow, which allows the membrane to continually clean itself. Only a portion of the liquid passes through the membrane. The rest continues downstream, sweeping the rejected materials away from the membrane to prevent any build-up. The main energy requirements for the reverse osmosis process are to drive a pressure pump to force the fluid through the membrane.

Reverse osmosis is capable of rejecting bacteria, salts, sugars, proteins, particles, dyes, and other constituents that have a molecular weight of greater than 150-250 daltons. The separation of ions with reverse osmosis is aided by charged particles. This means that dissolved ions that carry a charge, such as salts, are more likely to be rejected by the membrane than those that are not charged, such as organics. The larger the charge and the larger the particle, the more likely it will be rejected. Commonly reverse osmosis membranes reject from 96% to 99% of dissolved salts.

6.3.3.4   Operation of a Conventional Reverse Osmosis Unit

A conventional R.O unit consists of four major sub-systems:
- Pre-treatment
- Pressure Pump

- Membrane system
- Post treatment

Because of the membrane's sensitivity to fouling/clogging the feed water must first be pre-treated. This may be as simple as cartridge filtration in order to reduce turbidity or more complex, such as the addition of chemical anti-scalents, pH alteration etc. The water is then pressurised via a high-pressure pump and fed to the membrane(s). A throttle valve is used to hold the system pressure at its optimum operation point and regulate the proportion of water that flows through the membrane and that which will be exhausted removing the impurities. The ratio of the purified water to that of the feed is termed the Recovery Ratio.

Because of the necessary self-cleaning process, usual recovery ratios are approximately 10 –35%. This is inherently an energy inefficient process as the majority of the energy that has been input to the system is exhausted in the reject stream. Literature suggests that small R.O plants (up to 100L/h) use between 6-8 kWh per 1000L of purified water.

Post treatment is sometimes performed on the product water of a R.O plant. In small systems it may take the form of an activated carbon filter to polish the permeate and remove any odours that may be present. In commercial sized systems, pH correction may be implemented if necessary or a residual disinfectant may be injected into the product stream so that the water may be safely stored for a period of time.

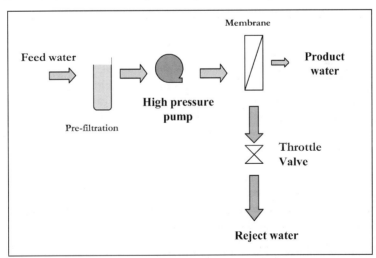

Figure 153:   *Schematic of a Conventional Reverse Osmosis Unit*

6.3.4   The Solar flow Technology

The need exists for a water treatment system suitable for remote area operation that is capable of treating a wide variety of water sources ultimately producing drinking water to WHO guidelines. The system needs to be:

- Robust
- Completely integrated, requiring no custom adaptations
- Easy to operate and maintain
- Automatic day to day operation
- Non-reliant on external power sources
- Easy to install and commission
- Cost effective / energy efficient

These criteria are mandatory, if a system is to be successfully operated in a remote location particularly in a developing country.

In the area of water supply and purification, appropriate technology can dig shallow wells, provide filtration of turbidity and reduce levels of micro-organisms, and may be able to store water from existing freshwater sources. What it can't do is effectively treat a polluted source for drinking water quality, or drill water wells beyond modest depths, or store water under sanitary conditions.

However, the technology to do these tasks does exist outside of the appropriate technology arena. State-of-the-art technology can be harnessed to develop drinking water supplies that far exceed appropriate technology capabilities and expectations.

The technology meets these criteria by blending proven, state-of-the-art technologies powered by solar energy to produce a low impact, sustainable solution.

6.3.4.1 Background

The desalination unit is a specially designed solar powered, reverse osmosis unit that has been aimed to meet the potable water needs of remote settlements and communities. The unit produces up to 400 litres per day, of high quality drinking water that exceeds the guidelines set out by the WHO and NH&MRC. The Solarflow removes dissolved salts, colours, heavy metals, pesticides, nutrients, viruses and bacteria (including cystic pathogens such as Giardia).

Development of the Solarflow is the culmination of approximately a decade of research and development. Investigations were initially performed in the area of renewable energy based water purification, by researchers from Murdoch University Environmental Science department.

This important preliminary research demonstrated that with an appropriate innovative design, a reverse osmosis plant could be built that was suitable for cost effective operation from photovoltaics. A local, West Australian, renewable energy company became involved offering engineering design expertise. Numerous different configuration prototypes were built and field tested in arduous working conditions within Australia and also internationally before the current design was selected.

6.3.4.2   The Solarflow Unit

The Solarflow's basic design philosophy was to produce a small, robust and cost effective water purifier that was capable of purifying a wide variety of feed waters and be completely autonomous in it is power supply and operation. Photovoltaics were the preferred power source because of their portability and low maintenance.

Reverse osmosis was chosen as the purification method because of its ability to remove not only pathogens but also dissolved salts, which enables the unit to be broad based in its application. Battery free operation was also another key design criteria as storage batteries are often the major cause of remote power system failures.

The engineering challenge lay in the fact that reverse osmosis is inherently a very energy intensive process, as only a fraction of the processed feed water is purified. This fact makes powering a conventional reverse osmosis plant from photovoltaics a costly exercise, due to the relative expense of the solar power system necessary.

Another draw back with conventional reverse osmosis plants, is they are designed to operate at a fixed speed (since they are designed to operate from a constant AC power supply) and any variation from this point not only reduces efficiency but may also lead to premature system failure.

Due to these criteria, it became obvious that a specially designed system would have to be produced. It must not only be very energy efficient, but also capable of operating effectively over a broad range of power input levels that are associated with photovoltaic generated power.

The Solarflow's unique, positive displacement pump incorporates a reject-water energy recovery system. This feature makes the Solarflow up to four times more energy efficient than conventional reverse osmosis units of comparable output. This highly efficient design not only enables the unit to be economically powered by photovoltaic panels but also enables battery-free operation.

A feature that further increases efficiency and reduces associated battery maintenance. The unit is fully automatic in its operation and can produce up to 400 litres of high quality drinking water per day from a single 100W solar panel.

Because the Solarflow is designed for use in remote areas, both low maintenance and ease of maintenance are paramount. All moving parts of the pump and energy recovery system have been incorporated within one assembly in order to minimise the amount of moving parts and importantly to simplify trouble-shooting.

If any problem is reported it then becomes a simple matter of replacing the pumping assembly and returning the defective pump for repair —field troubleshooting and repair is thus not required.

Changing the pump is a simple process needing the removal of two bolts and four hoses, which enables replacement by persons of limited technical skills. Similarly membrane and filter replacement procedures have also been designed for simplicity.

Figure 154: The Solarflow System

The reverse osmosis membrane used has been grossly oversized and the recovery ratio

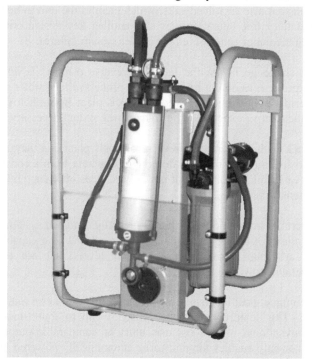

(feed water to product water) is relatively low in order to increase the operating life of the membrane. Average membrane life is approximately 3-5 years.

Chemical pre-treatment of the feed water is avoided because of the additional cost and complexity it involves. Since a single membrane is used, it is more cost effective to clean or change it on a periodic basis rather than introduce additional complexity of pre-treatment in an otherwise simple system.

Prior to this integration, Solarflow units were deployed using a wide variety of components making up the balance of system (power electronics, solar panels, water pumps, water tanks and tank stands). These components were selected and assembled on a customised basis for each application. It was identified that considerable benefits were attainable by optimising and standardising the system. These benefits would translate not only to superior performance but also increased customer confidence due to proven system integration. The Solarflow System was born out of this realisation.

The name "Solarflow System" refers to the complete autonomous water supply and purification system including:

- Solarflow reverse osmosis unit
- Sun Mill, solar powered water pump
- Solar panels and control system for the Solarflow and Sun Mill
- Water tanks
- Tank stands
- Pre-filtration system
- Potable water residual sanitation system.

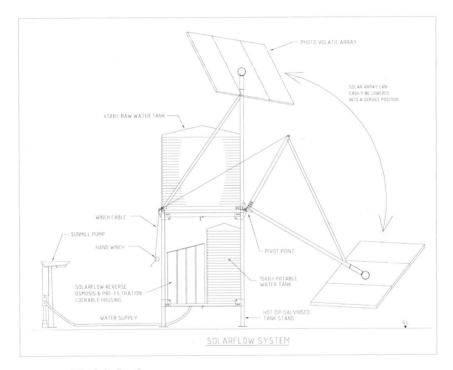

Figure 155: The Solarflow System

6.3.4.3   Economics of the System

The World Bank estimates that up to EUR 350 billion will be required in the next ten years to deliver clean water supplies to those currently without. As previously mentioned, up to 40% of the cost of delivering water in remote areas is in the energy needed to pump and purify supplies. The Solarflow System is a new, cost effective, environmentally friendly solution to combat this appalling domestic and international problem. There are four main alternatives for supplying water in remote locations. Diesel powered water purification systems are readily available. These usually take the common form of a diesel generator powering conventional water purification technology. Average small- scale systems are likely to prove excessively costly to operate. The bulk of this cost arises from the ongoing need to purchase diesel. Providing diesel to remote locations on a regular basis is often impractical and creates inflated prices. The regular maintenance issues relating to the operation of a diesel gen-set also incurs significant costs.

Transporting bulk water to remote locations is possible but often highly impractical and quite expensive. It is common however, for small amounts (a day or two supply) of water to be transported over long distances for family use. The WHO has reported that, annually, over ten-million person years of time are spent collecting water from distant sources, a staggering amount of time that could be spent in more productive activities if a viable alternative was available. Consuming contaminated water is the most common option. An option that results in millions of deaths annually, due to the fact that internationally, one in five people do not have access to safe and affordable drinking water (WHO, 1996).

Arriving at a cost incurred by a community that does not have access to safe drinking water is difficult to quantify, but the magnitude is enormous in both a financial and social sense.

Solar distillation is another alternative that has shown promise in some regions, but cost effectiveness is usually limited to small-scale systems – less than 20 litres per day.

6.3.4.4   The Solarflow System

Simple life cycle costing has been analysed which shows the projected lifetime cost of purified water. The analysis does not allow for depreciation, interest rates, or the rising cost of fossil fuels.

	Typical System Capital Cost (output approx. 60L/hr)	**Lifetime Cost of Water**
Diesel+ Reverse Osmosis System	EUR 12,000	3 - 4 c / litre
Conventional Solar +R.O System	EUR 30,000	2 c / litre
Solarflow System	EUR 20,000	1 c / litre

The Solarflow System fills the niche in this market, because it was designed to operate specifically from solar generated power, it is fully automated and low maintenance. Simple cost analysis shows it to be the most cost effective reverse osmosis unit for off-grid applications.

6.3.5   Conclusion

A huge demand exists for water purification technologies that are appropriate for developing communities and areas without grid electricity. In the past, projects have often been completed using an amalgamation of available technologies to produce systems which are theoretically capable of performing the task, but without the appropriate technology to ensure long term successful operation.

After a lengthy development period, the technology now exists to provide safe drinking water to isolated communities. The presented technology is sustainable, cost effective and straightforward to disseminate into developing regions around the world with proper funding programs.

6.4   Telecommunication

6.4.1   *Remote areas*

In remote and inaccessible areas, day and night, through all kinds of weather, solar power systems are used to keep communications channels open the world over. Reliability, low maintenance requirements and proven ability to provide continuous power in virtually all environments, have led to the use of solar power in telecommunications systems of every size and type. From very small aperture terminal (VSAT) phones to microwave relay stations, to television and radio transmitters - solar energy has been used as power source. As many telecom installations are installed in remote and unoccupied sites, such as a telecom repeater station on a mountaintop, diesel generator sets might be considered an expensive and unacceptable alternative.

Figure 156: Solar telecommunication installation in remote site

The regular supply of fuel might not only be a challenge in severe weather conditions but may also become a time intensive and expensive task. It is in these extreme circumstances that PV Systems have proven to be a reliable source of electricity. Although diesel systems have lower installation costs, the savings are quickly offset by the higher running and maintenance costs when compared to solar. After several years, the solar system installation might become the cheapest overall solution.

6.4.2   Typical solar-powered telecommunication system

The following diagram shows the components of a solar-powered telecommunications system.

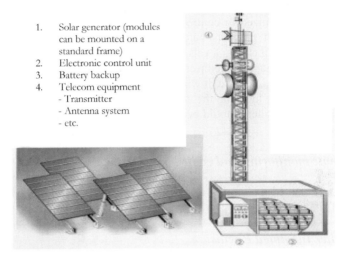

1. Solar generator (modules can be mounted on a standard frame)
2. Electronic control unit
3. Battery backup
4. Telecom equipment
   - Transmitter
   - Antenna system
   - etc.

Figure 157: Solar powered communication system

A typical telecommunications system installation includes the following major components:

6.4.3   Solar Array.

Is the full collection of all solar photovoltaic generators for a larger pumping system several dozens of PV modules are interconnected. They are mounted on ground installations using a simple frame that holds the modules at a fixed tilt angle towards the sun.

6.4.4   Batteries.

An electrochemical storage battery is used to store the electricity converted by the solar module. During the day, electricity from the module charges the storage battery. Batteries are typically 12-volt lead-acid ("car") batteries, ranging in capacity from 20-100 Amp-Hours (Ah). Batteries are typically sized to provide several days of electricity or "autonomy", in the event that overcast weather prevents recharging.

Deep-cycle batteries are best for a telecom system, as they are designed to operate over larger ranges of charge levels. While lead-acid batteries are only designed to be discharged 15% of their maximum charge, deep-cycle batteries can be discharged to 70-80% without incurring damage. Both deep-cycle and lead-acid batteries are typically used, as they are readily available throughout the developing world. Car batteries have a 3-5 year lifetime; deep-cycle, both sealed and unsealed, can last 7-10 years.

6.4.5 Charge Controller.

A charge controller is utilized to control the flow of electricity between the module, battery, and the loads. It prevents battery damage by ensuring that the battery is operating within its normal charge levels. If the charge level in the battery falls below a certain level, a "low voltage disconnect (LVD) will cut the current to the loads, to prevent further discharge. Likewise, it will also cut the current from the module in cases of overcharging. Indicator lights on the controller display the relative state of charge of the battery. Charge controllers should be appropriately sized to handle the peak power output of the solar array.

6.4.6 Remote monitoring and control system.

A remote and automated monitoring and control system is often installed to minimize the number of site visits.

6.4.7 Shelter/Insulated Enclosure.

The enclosure protects the sensitive electronic components from the elements.

6.4.8 Back-up diesel generator (optional).

In areas with marginal solar resources a back-up diesel generator might be installed. The usual PV power system would be relied upon for normal operation. Yet the diesel would be present if the battery bank became too depleted during a period of prolonged bad weather. The system controller could sense battery voltage, and turn on the diesel before a dangerously low voltage was reached. The generator would quickly recharge the battery bank, and then be turned off by the controller.

6.4.9 Telecom equipment.

Very small aperture terminal (VSAT) phones to microwave relay stations, to television and radio transmitters have been power by PV systems. Current state-of-the-art designs in telecommunications equipment require much less electrical power than a few years ago.

6.4.10 Application

6.4.10.1 Case 1
Solar Installation R.E. El Negrito—Telecom Argentina
Location: Tucumán, Argentina
Description: The repeater station at El Negrito is located at an elevation of 4200 meters above sea level and approximately 140 km from the city of San Miguel de Tucumán.

The inaccessibility of the mountainous terrain required pack animals to transport the equipment to the installation site. The system has a peak power output of 9,400 watts and allows more than 10 communities in the Calchaquies valley to be connected with the national communications system. Also, it allows the transmission of all communications traffic from the Providence of Salta to the Providence of Tucumán.
Climate: 20°C below zero with wind velocities reaching 120 km/hr.

Figure 158: Case 1,Telecommunication (left)   *Figure 159: Case 2 Telecommunication (right)*

System Components: One hundred eighty-eight 50-watt solar module, 48 volt system, with 4,000 amp-hours of battery capacity. Three system controllers diesel generators were installed for auxiliary power in case of prolonged bad weather.
Comments: The installation consumes 800 Watts of constant power. The solar array produces 9,400 peak Watts.
Cost: $180,000-$200,000 for power system materials and installation.

6.4.10.2   Case 2: Sun shines on Mexico's rural radio-telephony project

Location: Serving thousands of small communities throughout Mexico
Description: One of the largest exclusively solar powered communications systems in the world, Mexico's rural radio-telephony project has installed over 8,500 rural telephones and over 700 repeaters since 1989. The decision to use PV for all the radio and repeater stations provided the most economical and efficient implementation.
Climate: Varied
System Components: Cellular repeater stations: Twelve to twenty 48-watt solar modules with 1,200-2,000 amp-hour (Ah) capacity battery storage systems

Subscriber systems: Three to five 48-watt solar modules with 200-500 Ah capacity battery storage systems.
Cost: Repeaters: $10,000 to $40,000 for power system materials and installation.
Subscriber systems: $2,000 to $3,000 power system materials and installation.

Figure 160: Solar powered communication station in south Libya (El Bassam 2002)

6.5   Mobility and Alternative Fuel Vehicles

6.5.1   Traditional Draught Animals

- Power supplied by draught animals is the principal source of motive power in developing countries for small farms up to 80 – 90 % in the case of Africa and Asia
- 400 Million draught animals worldwide with a total installed capacity in excess of 100 GW, total energy supply is about 90 TWh or 320 PJ per year

Figure 161: A camel transports cooling box, for medicine powered by solar energy in the Sahara.

6.5.2 Alternative Vehicles

Alternative fuel vehicles, or AFVs, use alternative fuels instead of gasoline or diesel fuel. Currently, the majority of AFVs are sold as fleet vehicles.

Flex-Fuel Vehicles
Flex-fuel vehicles can be fuelled with gasoline or, depending on the vehicle, either methanol (M85) or ethanol (E85). The vehicles have one tank and can accept any mixture of gasoline and the alternative fuel.

Bifuel or Dual-Fuel Vehicles
Dual-fuel vehicles have two tanks - one for gasoline and one for either natural gas or propane, depending on the vehicle. The vehicles can switch between the two fuels.

Dedicated Vehicles
Dedicated vehicles are designed to be fuelled only with an alternative fuel. Electric vehicles are a special type of dedicated vehicle.

Alternative Fuel Heavy-Duty Trucks and Buses
Alternative fuel technologies have also been applied to heavy-duty trucks and buses.

Electric Vehicles
Electric vehicles have clear emissions benefits because they have no tailpipe emissions. They are perfect for driving limited ranges, and some models have achieved outstanding vehicle performance milestones while maintaining reasonable costs to purchase and operate the vehicles. Electric vehicles benefit from improved vehicle climate control technologies and advanced battery thermal management. Because of their limited use, however, the recharging infrastructure is sparse.

See the Components and Materials sections for information on fuel cells, high-energy batteries, electric drives, and lightweight materials—all of which are enabling technologies for EVs.

7   System Integration

The Food and Agriculture Organization (FAO) of the United Nations, in support of the Sustainable Rural Environment and Energy Network (SREN) has developed in cooperation with the Federal Research Centre for Agriculture (FAL), Germany a concept for the optimization, evaluation, and implementation of integrated renewable energy sources in rural communities.

7.1   The Integrated Energy Farm

7.1.1   Concept

The concept of an Integrated Renewable Energy Farm (IREF) (El Bassam, 1998) is a farming system model with an optimal energetic autonomy for food production in rural communities including options for energy exports. Energy production and consumption at the IREF has to be environment-friendly, sustainable and eventually based mainly on renewable energy sources. It includes a combination of different possibilities for non-polluting energy production, such as modern wind and solar electricity production, as well as the production of energy from biomass.

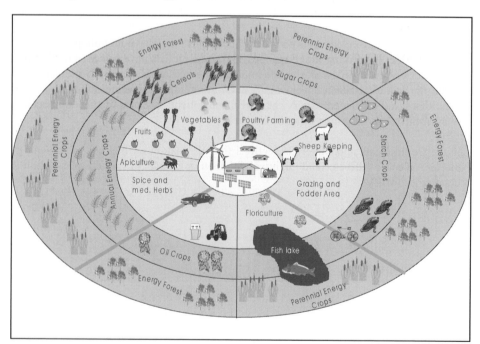

Figure 162: Model of an Integrated Renewable Energy Farm (IREF) (© El Bassam, 1998)

The IREF concept has been conducted for decentralized living areas, settlements and rural communities in which the daily necessities (food and energy) can be produced directly on-site with minimal external energy inputs. The IREFs will improve the systems of growing food crops, fruit trees, annual and perennial energy crops and short rotation forests, along with the development of wind, bioenergy and solar energy units within the settlements. This offers the possibilities to develop micro-enterprises and training sectors in order to improve the economic situation of the people.

An Integrated Renewable Energy Farming system based largely on renewable energy sources would seek to optimize energetic autonomy and ecologically semi-closed system while also providing socio-economic viability and giving due consideration to the newest concept of landscape and bio-diversity management. Ideally, it will promote the integration of different renewable energies, promote rural development and contribute to the reduction of greenhouse gas emission.

7.1.2   Global Approach

Basic data should be available for the verification of an IREF. Various climatic constraints, water availability, soil conditions, infrastructure, availability of skills and technology, population structure, flora and fauna, common agricultural practices and economic and administrative facilities in the region should be taken into consideration. From a global point of view predominating assumptions have been established in order to evaluate the possible contribution of the major renewable energy sources to support food production for a given area.

Climatic conditions prevailing in a particular region are the major determinants of agricultural production. In addition to that other factors like local and regional needs, availability of resources and other infrastructure facilities also determine the size and the product spectrum of the farmland. The same requisites also apply to an IREF. The climate fundamentally determines the selection of plant species and their cultivation intensity for energy production on the farm. Moreover, climate also influences the production of energy-mix (consisting of biomass, wind and solar energies) essentially at a given location; and how a single technology can be installed also depends decisively on climatic conditions of the locality in question. For example cultivation of biomass for power generation is not advisable in arid areas. Instead a larger share can be allocated to solar energy techniques in such areas. Likewise, coastal regions are ideal for wind power installations.

Taking these circumstances into account, a scenario was made for an energy farm of 100 ha into secific climatic regions of Northern and Central Europe, Southern Europe, Northern Africa and Sahara and Equatorial regions. It was presumed that one unit of this size needs about 200 mega-watt-hours (mwh) heat and 100 mwh power per annum for its successful operation. This calculated to a need for fuel of approximately 8,000 liters per annum. The possible shares of different renewable energies are presented in Table 34. It is evident that in all of Europe, wind and biomass energies contribute the major share to the energy-mix, while in North Africa and Sahara, the main emphasis obviously lies with solar and wind energies. Equatorial regions offer great possibilities for solar as well as biomass energies and little share is expected from wind sources of energy in these regions. Under these assumptions, in Southern Europe, Equatorial regions and North and Central Europe, a farm area of 4.8, 10 and 12 per cent of total land area, respectively, would be needed for cultivation of biomass for energy purposes. This would correspond to annual production of 36, 45 and 60 tons for respective regions. In North Africa and Sahara regions, in addition to wind and solar energy, 14 tons of biomass from 1.2 per cent of the total area would be necessary for energy provisions.

Moving ahead, in order to broaden the scope and seek the practical feasibility of such farms, the dependence of local inhabitants (end users) is to be integrated in this system. Roughly 500 (125 households) persons can be integrated with one farm unit. They have to be provided with food as well as energy.

Table 34: *The possible share of different renewable energies in diverse climatic zones producedon an energy farm of 100 ha area.*

Climate Region	Energy source	Power production (% Of total need)	Heat production (% of total need)	Biomass need [t/a]	Biomass area (% Of the total area)
North and Central Europe	Solar 200 m²	7	15	60	12
	Wind 100 kW	100	-		
	Biomass	100	105		
South Europe	Solar 250 m²	12.7	40	36	4,8
	Wind 100 kW	100	-		
	Biomass	70	65		
North Africa Sahara	Solar 300 m²	21	90	14	1,2
	Wind 100 kW	75	-		
	Biomass	25	25		
Equatorial region	Solar 200 m²	18.2	37.5	45	10
	Wind 100 kW	45	-		
	Biomass	70	80		

As a consequence, the estimated extra requirement of 1900 mwh of heat and 600 mwh of power has to be fulfilled from extra sources. Under the assumption that the share of wind and solar energy in the complete energy provision remains at the same standard as without the households, 450 tons of dry biomass is needed to be produced to fulfil this farm demand at a location in North and Central Europe. For the production of this quantity of biomass, 20 per cent of farm area is to be dedicated for cultivation. In Southern Europe and the Equator, 15 per cent of the land area should be made available for the provision of additional biomass.

7.1.3 Layout of an IREF

The concept of the IREFs includes 4 pathways:
1. Economic and social pathway
2. Energy pathway
3. Food pathway and
4. Environmental pathway.

In order to generate its maximum output, it is necessary to determine the most appropriate technologies, equipments and facilities to be located in the settlements, depending on the natural site conditions, infrastructure and the skills available or to be created in the region. These would probably include:
- Wind mills
- Solar collectors (thermal)
- Solar cells (PV)
- Briquette and pelleting machines for biomass
- Bio-gas production units
- Power generators (bio-fuels, wind and PV operated)
- Bio-oil extraction and purifying equipment
- Fermentation and distillation facilities for ethanol production
- Pyrolysis unit
- Stirling motors
- Water pumps (solar wind and bio-fuel operated)
- Vehicles (solar, bio-fuels operated or draught vehicles)
- Monitoring systems
- Cooking stoves
- Technologies for water supply, distribution and purification
- Food processing facilities
- Micro-enterprises
- Training units and services in agriculture, trade and business
- Maintenance working groups

The overall objective is to promote a complete sustainable development of rural areas and farming systems. Depending upon these objectives, the verification of demonstration integrated renewable energy farms in various regions of the world has to be prepared taking into account the following influential factors:
- Impact, influence and needs of climate, soil and crops
- Ratio of required food/bio-fuel production
- Input requirements for cultivation, energy balance and output: input ratio
- Equipment choices (wind, solar and biomass generation and conversion technology)

- Diversification of cultivation methods, technologies of water supply and water purifications
- Improvement of economical, social and environmental situation.
- The information collected from different regions would support a decision about the optimal farm size, its regional adaptation and solve other related constraints.

7.1.4   Regional Implementation

To verify the implementation of the IREF a practical sense at a regional level, taking into consideration the climatic and soil conditions, planning work has been started at Dedelstorf (Northern Germany). An area of 280 ha has been earmarked for this farm, which would satisfy the food and energy demands of 700 residents. For settlement purposes, old military buildings are being renovated. This project is expected to be completed in three years. The main elements of heat and power generation will be: solar generators and collectors, wind generator, biomass combined heat and power generator, stirling motor and biogas plant.

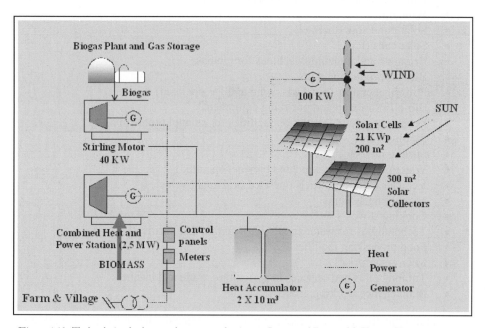

Figure 163: Technologies for heat and power production on Integrated Renewable Energy Farms

The total energy to be provided amounts to 8000 Mwh heat and 2000 Mwh of electricity. The cultivation of food and energy crops will be according to ecological guidelines. The energy plant species foreseen are: short rotation coppice, willow and poplar, miscanthus, polygonum, sweet and fibre sorghum, switch grass and reed canary grass and bamboo. Adequate food and fodder crops as well as animal husbandry is under implementation according to the needs of people and specific environmental conditions of the site. A research, training and demonstration centre will accompany this project.

In northern Germany a site near Hanover has been identified for the implementation of an Integrated Renewable Energy Farm and as a research centre for renewable energies – solar, wind and energy from biomass (90 % Biomass, 7 % Wind and 3 % Solar) as well as their configuration. Special emphasis is dedicated to optimisation of energetic autonomy in decentralised living areas and to promote regional resource management. The centre undertakes the responsibilities in the field of research, education, transfer of technology and co-operation with national and international organisations. It also offers the opportunity for trade and industry to introduce, demonstrate and commercialise their products. Co-operation with developing countries on the issues of sustainable energy and food production is also one of the prime objectives of the research centre.

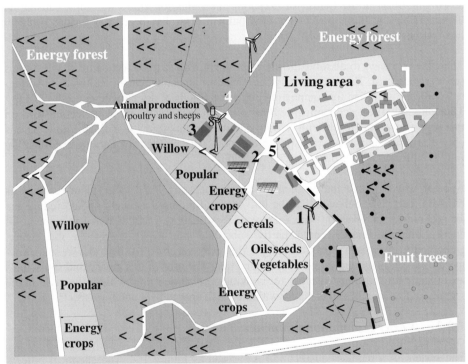

1 Thermal and Power unit (Biomass, Wind, Solar) 2 Pelleting, Oil mill, Ethanol unit
3 Animal husbandary 4 Biogas unit 5 Administration

Figure 164: Integrated Renewable Energy Farm, Dedelstorf, Germany (El Bassam 1999)

The importance of energy in agricultural, water and food production and consumption is evident and essential. An IREF represents an important step to achieve these targets.

7.2   New Achievement in the Utilization of Biomass for Energy in China (Q. Xi)

7.2.1   *Background*

China is a large agricultural country. In the countryside people have a long history of use of straws and other residues of plant materials as fuels for cooking and heating. But along with the accelerating development, mainly since 1980s, more and more coal and petrol products are consumed for these purposes, resulting in a surplus of mowed straws, remaining in the fields. The people prefer to use mineral fuels because they are easier to handle and cleaner and could be kept in the house. But in fact they are consumed at low efficiency, wasting natural resources, and causing severe air pollution. On the other hand, it is even more a waste of energy resources that the surplus straws are simply burnt in the fields, for preparing the fields for following crops, causing an unimaginable contamination of the environment. This has become a serious problem that is urgent to be resolved (Xi, 2001).

There are two ways to dissolve the above problem. One is to collect and take the surplus straws to compost. This is especially beneficial for an ecological agriculture. But a long-term process is needed for composting in the conditions of dry climate in north-China. There are few direct economical returns to be foreseen for the straw-collection work in this way of use. It is to difficult to persuade the farmers to do so and until now there is only a low percentage of straw put in compost. The problem of burning straw on the field is still serious or is becoming more serious.

Another way to solve the problem is to use the straws and other plant residues as biomass fuels. It is believed that the utilization of biomass as energy could reduce the air pollution in comparison with the use of fossil energy - and as renewable source of energy biomass energy could partly substitute the fossil energy in the future. Delightedly there is a breakthrough in this way recently. A new device of straw-thermogasifier is produced and put into practice. The uses of this device and straw fuels are suitable for the energy supply for cooking and heating in a modern life-style in the vicinity of a village. It can compete with the fossil energy in economic aspects.

7.2.2   *A brief explanation of the new device*

The system of a biomass gasification is showed as Figure 164. The chemical process taking place in the gasifier:

$$\text{biomass} \quad \xrightarrow[\text{heat}]{\text{imperfect combustion}} \quad CO + Co2 + H2 + CH4 + N2 + ash$$

The straws and other plant residues must be firstly cut into small pieces and then put in the gasifier through a transport band. The product gases are drawn with an electric fan (pump) through a gas-cleaner, where the ashes are filtered out, and then are stored in a large gas-holder. Finally the gases will go through connecting tubes to houses of consumers. The theoretical efficiency of energy production in this process is about 40%.

There are now four such biomass gasification stations constructed by Tianli Green Energy Corporation in 4 places in Yanzhou, Shandong Province. One of these is shown in Figure 165, Figure 166 and Figure 167. Investment of this station is about EUR 60,000. The designed capacity of gas supply is $1500 - 1800$ m³/day, to meet the demands of 300 consumer-houses. At present it has 100 consumers and burns corn-plant-residue or wheat straw of about 500 kg/day. The product gasprice is EUR $0.012 - 0.015$/m³ gas.

7.2.3 Perspectives

Although the uses of biomass fuel in a modern model is just at the beginning, it can be seen as the dawning of a new era of the utilization of biomass energy. The prices of the biomass-gases are now almost equal to or even lower than the petrol products in the countryside of China. On the base of economic reasons the use of biomass fuels with such devices should expaned quickly. This will improve the quality of the environment in large areas of the countryside and cover the deficits in energy resources now and in the future. In combination with biogas production through fermentation (which is mainly suitable in south-China), and new methods of composting - more progress must be achieved in biomass utilization, resolving the above mentioned problems, contributing to a sustainable agriculture, environment and energy supply.

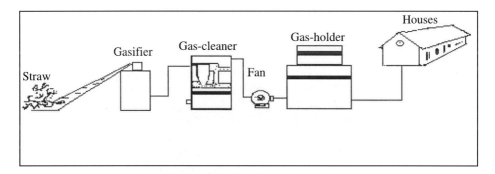

Figure 165: The system of biomass gasification (Xi 1998)

Figure 166: The gasifier and gas-cleaner (Xi 1998)

Figure 167: The gas-holder (Xi 1998)

Figure 168: A village supplied with biomass-gases (Xiaomaqing, Yanzhou, China) (Xi 1998)

8   Buildings and Energy Saving

Buildings are the modern world's most widespread technological systems and the most direct expression of a people's culture of life and work. Most of the energy we use – around 40% of primary energy in Europe – goes into heating, cooling and lighting building interiors and into running a growing number of devices used in buildings. Designing, building and managing energy-efficient buildings with low environmental impact is an ongoing challenge.

Over the past few decades, building roofs and walls have been continually transformed by the incorporation of new energy-related elements such as insulating materials to high-performance windows, special glass, solar-powered heating and electricity-generation systems, and low-consumption light bulbs.

Architects are switching to the "whole building" approach, which sees the various problems and solutions as a whole and tackles them in an integrated and intelligent way right from the start of the design process, when every choice is decisive.

The challenge is to move beyond the simple concept of "energy saving" or "solar energy" and aim at a combination of these and optimal building management system. The basic idea is to create better buildings by putting together a strongly interdisciplinary team capable of analysing and evaluating the different aspects involved in the building's life cycle, and striking a good balance among the proposed solutions. The factors involved include the building's site and position, and the use of active and passive solar systems.

The project must take into account waste management, maintenance, the choice and reuse of materials and products, optimisation of the technological installations, the financial aspects, the landscape and the environment, combining them all in an integrated whole.

The design process should dictate the choice of technologies, not the other way around, as often happens today, when available technologies and products guide the design process.

In recent years, the International Energy Agency's programmes on "Advanced Low-Energy Solar Buildings" have sponsored a number of products aimed primarily at energy saving and energy efficiency, but also at the introduction of solar technologies to meet the remainder of a building's energy requirements. These experiences have proved that it is possible to construct buildings that use on average only 44 kWh/m² per year, compared with 172 kWh/m² in other contemporary buildings. The lowest consumption obtained so far, 15 kWh/m², was in a home built in Berlin.

According to new building codes proposed in some northern European countries for future buildings, the amount of energy needed for winter heating can be reduced to practically zero with technologies that are already available (insulation, special glass, heat recovery, passive solar design and energy storage), and the remainder can be covered with active solar devices incorporated in the building's skin – devices that are not necessarily invisible, but are aesthetically designed for these buildings of the future.

9   Environmental Assessment

United Nation environmental conferences, United Nations Framework Convention on Climatic Change, UNFCCC and renewable energy organizations have highlighted the adverse local and global impacts of energy supply and end-use on the environment. The needs of sustainable development and the use of clean energy technologies are well recognized. Actions at national and international levels to tackle the problems of air quality and climate change are being developed, and these will require new directions for energy policy and technology and the nature of investment decisions in energy supply and end-use systems.

However the "Rio Declaration on Environment and Development" of 1992 did not recognize the intimate link between energy and development or the direct relation between conventional atomic and fossil energy supply and the vast majority of environmental problems:

- the threat to the atmosphere and the world climate;
- the increase of the ozone hole;
- the death of forests and the pollution of waters by acid rain;
- urban air pollution with fatal consequences for human health and the quality of life;
- the toxic contamination of seas, lakes and rivers;
- massive consumption of scarce water reserves in petroleum and coal production and processing;
- the risks of nuclear radiation and unresolved waste storage problems, burdening human civilization for thousands of years to come.

Figure 169:   Conventional fossil fuel centralized power plants.

Problems of bio-degradation, deforestation and other vegetation loss over large land surfaces are linked to dysfunctional energy consumption patterns, because large quantities of biomass used for energy demands are not renewed. Therefore, under the conditions of nuclear and fossil energy use (as well as not renewed biomass) the target of a "sustainable development" cannot be reached. None of these pending and visible dangers can be overcome without substituting nuclear and fossil energy with Renewable Energies. Energy projects in developing countries have the opportunity to benefit from global environment programmes, and to participate in renewable energy development mechanisms. These instruments offer the prospect of further international cooperation in rural energy development in order to achieve environmental, social and economic goals.

Renewable energy can play a major role in furthering a wide range of EU and international policy goals. In 2003 the German parliament took the initiative of promoting an International Renewable Energy Agency (IRENA) as an international governmental organization. Its purpose is to support and advance the active utilisation of renewable energies on a global scale. For 50 years has existed an international agency for the promotion of atomic energy whereas furthering of renewable energies irrespective of their worldwide potential has been more random.

These energies can, more than any other technologies, play a key role in the protection of the global climate and the environment, resource conservation, fight against poverty in developing countries and secure the long-term availability of supplies. An International Renewable Energy Agency can obtain the status of promoting worldwide technology transfer in connection with renewable energies and increasing energy efficiency not least efficient and integrated utilisation of energy.

Figure 170: Renewable energy training of technicians: Trainee from Ghana at the Folkecenter, Denmark (left); Kyocera Photovoltaic Training Centre in Sakura, Japan (right); Rural Solar Electrician School, Mali Folkecenter, Tabakoro, Mali (bottom)

The task of an international agency will lead to the concentration on the subject within the framework of a separate governmental organisation. The promotion of renewable energies will not only be a question of commercial energy supplies, but in many cases one of autonomous energy utilisation and the introduction of corresponding technologies. Since renewable energies open up the option of distributed energy utilisation in order to exploit its potential, it can become of crucial importance to broaden the knowledge basis in a variety of different occupational fields.

Development of a successful renewable energy sector would in the future make a useful long-term contribution to diversity, security and self-sufficiency of energy supply because they, in contrast to the fossil fuels, will not be depleted. Both in the industrialized countries and in the unserved areas with limited or no access to commercial energy supply, the transition to the renewable energies will create new, labour intensive, industrial sectors. The local and national economies will become less vulnerable to international conflicts and price fluctuations of fossil fuels.

In addition, as elimination of environmental impact is one of their main benefits, renewables could play a leading role in mitigating the environmental effects of energy use, since all the technologies covered in this book in various proportions offer major reductions in harmful emissions compared with fossil fuels (Figure 170).

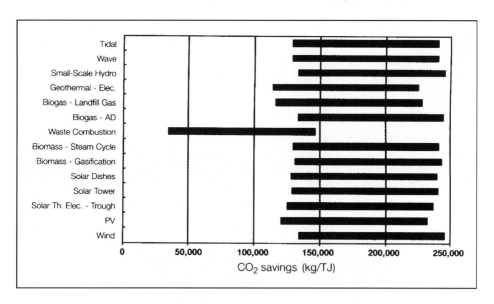

Figure 171: Saving in CO_2 emissions compared with conventional power generation

Figure 172: Integrated energy farms in Thy, Denmark: Wind energy and biogas (left); rape seed for PPO combined with wind energy (right).

The concept of "integrated energy farms" is based on the principle of an economy in synchronisation with nature - producing energy while protecting the environment, land and resources. The operation of the energy farm is climate neutral because here the energy supply of the site will be secured 100% by using renewable resources (sun, wind and biomass).

The exploitation of solar energy for heating and electricity supply occurs completely emission-free. Also the production processes of solar cells and collectors do not cause great environmental effects, if any.

The installation of windmills normally involves a significant visual impact on the local landscape. However, its environmental effect is very low compared with other available systems of energy supply. The noise emission of modern windmills, compared to former models, has been considerably reduced through aerodynamic improvements and noise isolation of mechanical equipment. Depending on the type of turbine the noise level directly at the turbine amounts to 60 to 100 decibels.

This, however, is not the sound level that residents living in the proximity of single windmills or a windfarm will experience. Governments in several countries are applying environmental standards with regard to noise emission. The example of Denmark where 6,000 windmills (2003) and an installed capacity of 3,200 MW cover 20% of the demand for electricity, environmental legislation states that noise at the closest resident must not exceed 45 dB(A). Within more densely populated areas the sound level must be maximum 40 dB(A), similar to the sound of subdued voices or whispering. Because a windmill by law cannot be sited closer than 4 times the height of the windmill to a residential building the above mentioned sound level will in practice never be attained, so at residential areas, a modern windmill is virtually non-audible.

Figure 173: Windmills placed close to residential and industrial buildings.
They comply with official noise standards of maximum 45 d(B)A.

Long-standing observations have shown that during the day birds in flight avoid wind turbines. Due to the windwills' extensive solid mass bird impacts are rarely observed. It is extremely low compared to bird collisions with high-voltage power lines, road traffic etc. However, one can in special regions observe that on dark nights and also during fog, migrating birds do occasionally strike the towers and blades of wind turbines. Therefore, wind power plants should not be installed in the main migratory routes or in protected nature areas.

The special advantage of the wind power plants lies in their CO_2-neutral electricity production. Compared to power generation using fossil energy sources with every kilowatt-hour of wind electricity, 700 g CO_2, 1 g SO_2 and 1.4 g NO_X can be avoided. In the case of a 1.5 MW turbine, 2,400 tons CO_2 per year and 48,000 tons CO_2 during its lifetime can be eliminated.

During its lifetime an efficient windmill installed at a windy site will produce 80-100 times the quantity of energy that was required to manufactur and install the windmill. Life cycle analyses illustrate that wind power production has less negative environmental impact than all other forms of electricity production. In only 3-4 months it delivers the amount of energy that in total was needed for its fabrication, installation, maintenance and final decommissioning.

Biomass, a photosynthesis product of CO_2 and H_2O, is one of the most important renewable energy sources. Actually, wood, charcoal and other woody materials play a dominant role in the traditional energy supply of many developing countries. During the energetic combustion of biomass, the same quantity of CO_2 is emitted which has been extracted from the atmosphere by the plants during their growth. Compared to the fossil energy carriers the energetic use of biomass does not cause any kind of additional load on the atmosphere by the CO_2-emission.

In Tables 35– 38 some characteristics of biomass emission are presented.

Table 35: *CO$_2$ - emission of different energy sources*
Source www.nerl.gov

	Dry matter yield	Heating value	CO$_2$ emission
	t/ha	kWh /kg	t CO$_2$ /t
Wood and other woody materials	12	4.6 (moisture 18%)	0.75
Bio fuel	1.32	11.13	0.74
Mineral oil		12.75	3.80
Fuel oil		12.81	3.80
Mineral coal		8.67	3.00
Brown coal		2.43	0.93

Table 36: *Energy data and CO$_2$ - emission of biogas*
Source www.nerl.gov

Type of animals	No.	cattle unit	gas content	fuel oil equivalent	gas produc-tion	methane content	CO$_2$ reduc-tion
			kWh/ m^3	l / k	m^3 / a	%	t/a
dairy cattle	100	100	6.0-7.0	27,000	45,000	60-70	100
sheep	100	40	6.0-7.0	13,200	22,000	60-70	50
piglets	2000	120	6.0-7.0	33,000	55,000	60-70	120
pigs	400	60	6.0-7.0	16,200	27,000	60-70	60
chickens	10,000	90	6.0-7.0	33,000	55,000	60-70	120
calves	500	90	6.0-7.0	33,000	55,000	60-70	120
young cattle	200	140	6.0-7.0	46,000	77,000	60-70	170

Table 37:   Energy data and CO_2 - emission of different solid biomass combustibles
Source www.nerl.gov

Type of biomass	moisture	weight	heating value	fuel oil equivalence	CO_2 – reduction potential	ash content
	%	kg / m³	kWh/kg	t / 1000 l oil	per t biomass	%
Straw	10		4.0	2.5	1.48	6
Energy plants	15	250	4.3	2.3	1.61	1
Bark beech /pine	15	200	3.8	2.6	1.42	9
Cereals	15	500	4.0	2.5	1.48	3
Beech /oak chopped wood (dry)	10	308	4.5	2.2	1.68	1
Beech /oak chopped wood (moist)	40	425	3.0	3.3	2.49	1
Pine/fir chopped wood (dry)	10	230	4.2	2.4	1.54	1
Pine/fir chopped wood (moist)	40	272	2.9	3.4	1.01	1

Table 38: CO_2 - *emission of an integrated energy farm compared to the conventional energy supply*
Source www.nerl.gov

	Integrated Energy Farm A	Conventional energy supply B
Source of energy	Biomass (chopped wood dry)	Fuel oil
Quantity	24. 0 t	10,000 l
Energy production	100 MWh	100 MWh
CO_2 - emission	18.0 t	31.92 t
CO_2 - emission (A-B)	**- 13.92 t**	

Figure 173 provides important information on the rate of fossil fuel consumption, CO_2 concentration in the atmosphere, air temperature and the cost of storm damages in the period from 1970 to 2000. This reflects very clearly the close correlation between fossil fuel utilization and other parameters.

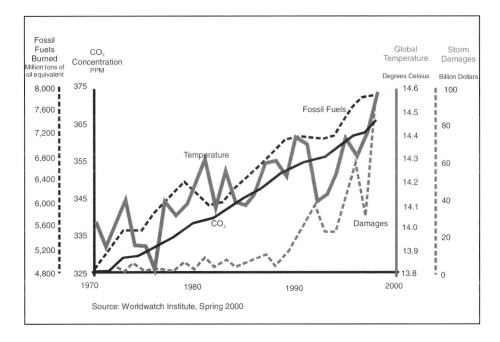

Figure 174: Energy and the climate

10 Economic Dimensions

With the exception of nuclear, geothermal and tidal energy, all forms of energy used on earth originate from the sun's energy.

Renewable is the term used for forms of energy that can be regenerated, or renewed, in a relatively short amount of time. The regeneration process may be continuous and immediate, as in the case of direct solar radiation, or it may take some hours, months or years.

This is the case of wind energy (generated by the uneven heating of air masses), hydro energy (related to the sun-powered cycle of water evaporation and rain), biomass energy (stored in plants through photosynthesis), and the energy contained in marine currents.

The energy contained in fossil fuels – coal, oil and natural gas – likewise comes from the sun's energy, but it was stored in plants millions of years ago, and once used, it cannot be regenerated on a human time scale. The earth's remaining fossil fuel reserves can probably provide us with energy for another 100 to 500 years, but this is an insignificant amount of time in terms of the whole past history of human civilisation and (one hopes) of its future.

The flow of renewable solar energies on earth is essentially equal to the flow of energy due to solar radiation. Every year, the sun irradiates the earth's landmasses with the equivalent of 19 trillion toe. A fraction of this energy could satisfy the world's energy requirements, around 9 billion toe per year.

10.1 Economic Aspects

For economic efficiency in the use of renewable energies, the energy production costs are very important. The total energy cost is determined from different cost data by use of a simple method (annuity method). Generally the energy cost is calculated as follows:

$$\frac{\text{total annual cost (depreciation + operating cost)}}{\text{total annual energy output}}$$

For the solar energy generation plants the energy cost depends on the capacity and is calculated as follows:

$$K_a = [q^{n}{*} + (q-1) / (q^{n}{*} - 1) + Z_B / 100]\, k\, \text{Euro} / \text{KWh}$$

$$q = 1 + p_z / 100$$

K_a : specific annual cost (Euro / KWh)
p_z : interest rate (% /a)
Z_B : operating cost rate (% /a)
k : specific cost of plants (Euro / KW)
$n{*}$: service life (a)

The electricity production cost will be:

$$K_s = K_a / n_v (\text{EUR}/\text{KWh})$$

K_s: electricity production cost (EUR/KWh)
n_v: full load hours per year (h/y)

For facilities using biomass (wood, crop waste etc.) the energy costs can be also calculated by using the formula mentioned above. The particular facility cost normally depends on the size and capacity of the facility as well as on the available quantity and type of biomass.

For the use of cogeneration systems, which produce simultaneously both heat and electricity, the economic efficiency of a facility depends on the exploitation degree of produced energy. For example, if biogas is exploited on a farm for space heating and for power generation, a part of the biogas will be not used optimally in the summer, due to the fact that at this time we generally do not need much heat energy.

The integrated energetic use of different renewable resources will be economically more favourable if a constant energy demand exists on the site for both heat energy and electric power.

Figure 175: Advanced renewable energy technology: Biogas plant by Kawasaki Engineering in Yubetsu, Japan (left); combined heat and power, CHP by Caterpillar (middle); and windfarm by Vestas (right) photo V.Kantor.

The efficiency of wind turbines is different from site to site. The wind resource and the generation capacity of turbines play a big role. For the calculation of the wind energy production at a specific site one can utilise either wind measurements, or the wind atlas method applying with the land surface roughness as the main parameter. Taking the extremes the same windmill with the same investment will have an annual power production that is of a factor three higher at open land sites without trees, hills or buildings compared to a site with considerable obstacles in the landscape that are influencing the flow of the wind. Electricity can normally be produced in an economically favourable way by using windmills having capacities of more than 200 kW and on sites with a average wind speed higher than 4.5 m/sec.

Figure 176: 4x525 kW windmills each with annual production of 1,400,000 kWh. Built 1992/93. Situated on the beach of the North Sea, near industrial complex and fishing harbour of Hanstholm, Denmark

In the table 39, some standard values are given for energy costs under the environmental conditions of Northern Germany. It indicates the costs per kWh of electricity produced by renewable energies. However, the external costs of the pollution caused by conventional power generation are not usually considered by individual investors, and the renewables typically offer more to society and more to their users than simply an energy service

For the planning of integrated energy farms it is very important to calculate first the economical feasibility of energy plants by using available and measured necessary data. This will be very different from site to site. A general approach applicable everywhere does not exist.

Table 39:   Production cost of heat and power from different energy resources

Source of energy	electricity	heat energy
solar	0.60 –1.00 EUR / kWh	0.05 – 0.08 EUR / kWh
wind	0.05 – 0.12 EUR / kWh	-
biomass	0.04 – 0.08 EUR / kWh	0.01 – 0.02 EUR / kWh

Figure 177: Historical and future costs of electricity produced by renewable energy technologies

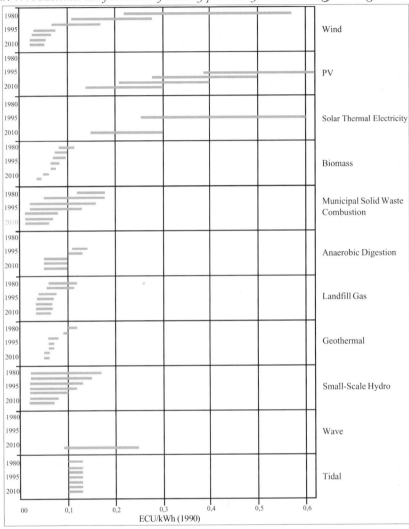

ECU/kWh (1990)

10.2 Recent and Future Market

Hydropower accounts for 19% of the world electricity supply, utilising one third of its economically exploitable potential. Hydro projects have the advantage of avoiding emissions of greenhouse gases, SO_2 and particulates. Their social impacts, such as land transformation, displacement of people, and impacts on fauna, flora, sedimentation and water quality can be mitigated by taking appropriate steps early in the planning process. Whilst a question remains over the advantage of smaller hydro schemes over larger ones (owing to the farmer's greater total reservoir area requirement), it is believed that generally hydropower is competitive, when all factors are taken into account.

The production of Wood fuels is estimated to cover nearly 6% of the world energy requirement, although there are undoubtedly some difficulties in quantification. Wood fuels' share is thus larger than that of hydro and other renewable energy resources, but smaller than that of nuclear. A re-evaluation of wood fuels has shown that considerable amounts are now estimated to come from non-forest sources. Wood fuels continue to be used traditionally in rural areas of developing countries where they remain a burden for women and children to collect and, owing to their incomplete and inefficient combustion, also hazardous to health.

Whilst rising income levels and urbanisation in developing countries have resulted in a reduced share of wood fuels in their overall energy use, changes in energy and environmental policies, such as global warming mitigation in developed countries, has led to an increased use of wood fuels, often as modern biomass.

A special Report on Emissions Scenarios produced by the Intergovernmental Panel on Climate Change (IPCC) has concluded that although the longer-term maximum technical energy potential of biomass could be large (around 2,600 EJ), this potential is constrained by competing agricultural demands for food production, low productivity in biomass production, etc.

Despite a growing interest in biomass (other than wood), as a result of energy market reforms, environmental concerns, and technological advances, the major remaining challenges are the low combustion efficiency and health hazards associated with traditional use of bio-energy. Because of the many difficulties in assessing the energy potential of residues, it is suggested that the focus should be on the most successful forms such as sugar cane bagasse in agriculture, pulp and paper residues in forestry and manure in livestock residues.

The modernisation of biomass use relates to a range of technological options, such as gasification, co-firing with fossil fuels, micro-power, tri-generation, and ethanol. It is argued that biomass can directly substitute fossil fuels, as more effective in decreasing atmospheric CO2 than carbon sequestration in trees. The Kyoto Protocol encourages increased use of biomass energy.

Raising the contribution of solar and other renewable resources to 50% of world energy use by 2050, as suggested in the Shell Renewables report, would require sweeping changes in the energy infrastructure, a new approach to the environment and the way that energy is generated and used. Despite the development of modern solar energy over the past thirty to forty years, the technology still needs a higher profile and more involvement from scientists, engineers, environmentalists, entrepreneurs, financial experts, publishers, architects, politicians and civil servants. A new generation of solar-energy pioneers has to be nurtured. Appropriate policies and strategies for the proliferation of renewable energies are also required.

Since the 1998 Survey there has been an increase in world geothermal plant capacity and utilisation, for both power generation and direct heat supply, but the pace of growth in power generation has slowed compared to the past, while that of direct heat uses has accelerated. Over-exploitation of the giant Geysers steam field has caused a decline in geothermal capacity in the USA in recent years, which has been partly offset by important capacity additions in other countries.

A large increase in the number of geothermal (ground-source) heat pumps has contributed to the increase in direct heat application. Although the short- to medium-term future of geothermal energy looks encouraging, its long-range prospects depend on the technological and economic viability of rock heat (HDR).

There has been a steady growth in the size and output of wind turbines, now available with capacities of up to 3 MW for offshore machines. The support provided by national governments influences development patterns: for example, contrast wind farms in the USA and the United Kingdom and single machines (or clusters of two or three) in Denmark and Germany. Environmental issues surrounding wind energy pertain to noise, television and radio interference, danger to birds, and visual effects, but in most cases, sensitive siting can solve these problems. Many utility studies have indicated that wind can be readily absorbed in an integrated power network until its share reaches 20% of maximum demand.

It is expected that due to the rapid capacity growth in many countries and regions, global installed wind capacity may reach 150 GW by 2010, depending on political support, both nationally and internationally, and further improvements in performance and costs.

The energy dimension is not the only one. Effects on development, trade and employment should also be considered. In fact, renewable energy development will contribute to poverty alleviation and to the growth of modern industry. Renewable energy technologies could also contribute positively to the balance of trade. However, many countries continue to subsidise fossil fuels, instead of developing the sources and supply of renewable energy.

The renewable energy sector is currently the fastest growing energy sector worldwide. In just two decades the wind energy industry has evolved from a collection of backyard enthusiasts to a multi-billion dollar industry experiencing a twenty to thirty percent annual growth rate in the 1990'ies.

The solar cell market has also experienced substantial growth, increasing by an average of 15 percent annually from 1990 to 2000. In 2003, the industry sold more than 400 megawatts of solar cell modules for a total revenue of more than EUR 3 billion.

This growth has been achieved by two complimentary forces; rapid technical progress from the experience of installing hundreds of devices has resulted in dramatic declines in the cost of wind energy. In the best sites, for example, wind is now competitive with most fossil fuel technologies.

Progressive legislation with guaranteed long-term tariffs such as the German Renewable energy law, EEG, the US state of Texas, the Australian renewable energy targets, and the Danish legislation with the purpose to reduce oil dependency in the period 1978-2002, are making it easier to bring new technologies to market.

Designation of sites and combined with appropriate policies have resulted in a number of wind energy prospectors creating a wind rush in especially Germany, Spain, USA, India and several other countries. This in turn is creating new markets for wind energy in shallow offshore areas.

10.3 Financing and Investment

Although these developments currently attract only a small fraction of the annual investment in new energy projects, the growing experience is giving many lenders and investors the confidence to invest in the renewable energy sector through various financing pathways:

Debt
Loan obligations usually as corporate finance or non-recourse project finance.

Private Equity (Funds and Fund Managers)
Capital invested in a private company by individual investors or firms that invest the pooled funds of investors for a fee.

Funds Invested in Publicly Traded Companies
Funds invested in company shares on public stock exchanges, or funds invested in other funds.

Export Credits
Trade credits provided by government agencies to companies in order to facilitate the financing of exports.

Insurance
Financial products that promise compensation for specific potential future losses in exchange for a periodic payment.

Funds and Financing to Serve (Technology Type):

All Energy Efficiency (EE) Types

EE technologies are generally used to reduce consumption by end-users, and so are called demand side technologies. They bring about environmental benefits by reducing the demand for energy. Examples include high efficiency motors and lighting, district heating and heat/electricity co-generation systems, heat pumps, and building insulation.

All Renewable Energy (RE) Types

RE technologies use non-depleting sources of energy, such as the sun or wind. RE technologies can provide either electricity or heat. RE are supply side technologies in that they supply energy. Those that generate electricity can either be used on-grid, thereby offsetting energy produced from conventional sources, or off-grid, to provide power in remote locations. All of the specific RE types listed below are included in this category.

10.4 Development

The two common factors that underlie many of the problems threatening our future are the fast growing population and the ever-increasing consumption of resources driven by the diffusion of life-styles that have developed in industrialised societies and are emulated in much of the world.

Until the discovery of fossil fuels and the beginning of the industrial revolution, the sun's energy – in its different forms, direct and indirect (such as wind and biomass) – was the sole energy source that inspired and enabled the development of human societies.

Since then, and especially in the past one hundred years - a relatively short span of time - a powerful energy infrastructure has been built that now covers practically the entire planet and is based on fossil fuels and nuclear energy has been built. Today the world consumes 9 billion toe per year, compared with around 500 million toe in 1860. While these energy uses and infrastructure do not yet benefit billions of poor people who still try to make do with firewood, they give humanity a power over nature that earlier generations never knew; they had to survive with the renewable energy of the sun.

This power helps us live more comfortably than past generations, but while it meets new needs, it also carries the risk of irreversibly altering natural balances, both local and global.

The world's population has been growing rapidly over the last century and continues to grow. We were 1.6 billion in 1900 and we have now passed the 6 billion mark. If this trend continues, the human population will rise to about 9 billion by 2050. The increasingly crowded world has also become a world of cities. Fifty percent of the population live in cities and the figure is expected to rise to 75% by the year 2050. Dozens of cities already number more than 10 million people.

Dramatic contrast between wealth and poverty has become part of any urban landscape, with excessive consumption among the richer segments, the inability of the poorer segments, especially in the developing countries, to meet their most basic needs: decent homes, clean water, health care, education. If these legitimate and ever-growing needs are to be met, energy consumption must increase.

A great number of solar, wind and biomass technologies for the production of fuel, heat and electricity are now available or close to commercialisation. They have been installed on a significant scale in both developed and developing countries. They are used in many different ways, stand-alone or incorporated in conventional energy networks and grids. They are already providing energy services to individual homes, villages and cities.

However, if we are to move from examples to worldwide applications of renewable energy technologies in communities, cities, islands and rural areas, society as a whole must be interested and provide its support. Solar energy infrastructure, whether installed in remote rural areas in a developing country or integrated in existing conventional infrastructure in a city in the developed world, needs to be better known and accepted.

If we want the use of renewable energy to spread through the technologically advanced world to the extent mentioned above – 50% of world energy consumption by 2050 – we will need to enroll many more renewable energy scientists and engineers, environmental scientists, entrepreneurs, financial experts, publicists and architects. Above all, we will need many more politicians and civil servants who know the subject and are more courageous and determined. A new generation of solar-energy pioneers has to be nurtured, especially to work in local communities and industries.

Solar energy exists everywhere, but has a weaker concentration or density of energy than fossil and nuclear sources. Using solar energy can teach us how to establish a more balanced relationship with nature. A new culture of energy efficiency can lead to a more concerned, socially responsible use of all natural resources. The use of solar energy – a local resource – can contribute to the preservation of local cultures and also promote new lifestyles and new concepts of wealth, prosperity and security that can help us all meet the challenges of the 21st century.

10.5 Marketing Strategy

New Paradigm

- Selling Electricity - or the system
- Creative Product Development
- Commercially driven distribution system
- Creating affordable financing scheme
- Quick response to technical problems solving - after sales services types of response
- Forming business partnership with rural based institutions

The Market Prospects

- Over 400 000 households in rural areas not reachable by PLN Grid

- Growing awareness among urban population for the need of "reliable" and renewable energy
- Global thrust reducing CO_2 emission and increasing awareness of depleting fossil based energy – Harare Declaration 1996 → Kyoto Protocol
- Higher Efficiency and Lower Cost of Solar Systems
- Gradual lifting of government subsidies on fuel oil makes PV systems increasingly competitive.

The Market Challenges (in Rural Market)

- Huge needs but low purchasing power
- Income are sensitive to global commodity price fluctuation ↔ exchange rates
- Lack of interest among financial institution in micro credit scheme
- Low knowledge of the characteristics of renewable energy systems ↔ prevailing expectation from promised 'grid expansion'
- Traumatic experience of using PV systems distributed free of charge but not followed by maintenance services and disturbing trade in 'used' systems with no quality assurance.

The Urban Market Challenges

- Protection against the impact of grid breakdowns
- Protection against the ever increasing price of electricity due to the diminishing fuel price subsidies
- Increasing awareness of the need of renewable and environment friendly energy
- Creating renewable energy system applications that meet typical urban standards of energy need.

The Market Challenges

- DEMAND:
  - o **Rural Areas** – Huge needs of *'basic electricity'* (Some 2 billions people having no access to electricity)
  - o **Urban Areas** – The increasing needs of *"reliable electricity"*
- CAPABILITIES
    Technology Progress: higher efficiency and lower cost, but slow in reaching the market
- GLOBAL SETTING
  - o Renewable Energy Thrusts and Imperatives
  - o Current Global PV production may not meet the rapid increase of global market demand
  - o Doubts following the US refusal to follow Kyoto protocol

Product Development (Response to Challenges)

Creative Product Development

Renewable Energy business should be committed to satisfy demand of rapidly growing markets, by creating systems that meet the specific requirements of its customers through strategic alliances with relevant industrial partners.

Competitive Factors in Product Development

- Environmental considerations
- Performance Reliability
- Emergency Services
- Peripheral utilization
- Stand-by applications
- Hybrid Systems

Market Experience (Some Lessons Learned)

- Identification of the market needs is relatively easy but evaluating its ability to consume the services offered is much more difficult.
- The srong family and communalties require unconventional marketing approaches to convince customers that RE is one of the best alternatives to grid electricity.
- Ability to provide the right financing scheme is the key to success in marketing PV in rural areas, where most of the prospective customers are poor or have an irregular stream of income.
- Ability to provide continuous services near the users' location is imperative to warrant sustainable business success.
- Quick response to the 'demand' of prospective customers is required - as in many occasions they were disappointed with the 'non-performing' agencies promising services to them.
- In distributing SHS, information on the individual system performances is needed to insure the continuity of effective services to the users.

Summary & Conclusion Remarks

- The market of RE Electricity is still in its infancy, hence it promises great prospects for the future.
- Effective and innovative distribution systems is an imperative key to success in marketing RE systems.
- Serving a vast but poor rural market definitely needs suitable and dedicated credit systems. Current Lack of interest among local financial institutions in micro credit systems forces RE companies to use their own financial resources to finance the credits given to the PV rural users.
- Dedicated product development efforts should be carried out to enhance the market and marketability of SHSs and other RE systems.
- Product development should look carefully into product improvements by incorporating local components produced by partner companies through strategic alliances.

- Marketing 'RE Electricity' requires the establishment of reliable RE systems, hence, the need for gaining consumers' confidence in PV system is very profound.
- Field staffs with adequate technical skills are required to provide effective services to RE users.
- Partnership with PLN as the authorized distributor of electricity in the country to achieve a win-win solution for all the stakeholders concerned.

11 Legislative Structures

Policies and strategies supporting the transition from the dominance of fossil fuels and atomic energy have in all countries decisive influence on the implementation of renewable and decentralized forms of energy. This can be illustrated by the successful elements of the 1991 German legislation, "Act on Feeding in Electricity". It resulted in a breakthrough for wind power achieving 50% of the total European capacity as well as progress for other types of renewables.

The initial legislation formed the background of the essential principles of the Renewable Energy Law, EEG, from 2000. However, the tariff structures are differentiated allowing specific renewable energy technologies a compensation determined by the stimulus which is required to maximize the implementation of a variety of energy solutions for the future.

New Industrial Development and Job Creation.
Increased use of renewable energy will create jobs, especially in the sector of small and medium-sized enterprises. They play a crucial role in the economic structure of most European countries. Small and medium-sized enterprises are not only an important factor in crafts and trades; they also provide an impetus for a variety of industries, including the metal industry, electrical engineering, mechanical engineering, biochemistry, as well as the building industry.

The stimulation of the use of biomass for electricity generation will also provide a major impetus for the agricultural sector. Furthermore, the production and use of renewable energy sources will promote sustainable regional development, which will help to improve the social and economic cohesion and to harmonise living conditions.

In three European countries – Germany, Denmark and Spain – (Denmark to 1st January 2000) national legislation has been adopted to introduce minimum prices for feeding electricity generated from renewable energy sources into the grid. It is exclusively due to the national legislation of these three countries that the European Union accomplished the emergence of a large wind turbine manufacturing industry. It provides the leading technology in the world market.

This also confirmed it was wrong to assume that the introduction of minimum price systems would hamper productivity. In all the three countries mentioned above the introduction of wind energy was based on minimum prices guaranteed by law. Market development was stimulated – initially in the wind energy sector – resulting in an efficient industry with considerable export opportunities, creating jobs for over 40,000 people in Germany and 20,000 in Denmark.

As a result of the associated economies of scale and the global competition among manufacturers of windmills, production costs have been successfully reduced by 50 percent since 1991. Owing to technological progress, there is growing demand in the world market. In the period to 2010, demand for windmills alone may amount to over 100,000 MW.

Against this background, the market introduction of renewable energy sources should not be underestimated in terms of its importance for industrial policy. It can be assumed in view of global climate problems, oil depletion etc. that there will be a rapidly growing demand worldwide. It can be expected that the impact on other renewable energy sectors, will be similar to the effects experienced on the wind energy sector.

Innovative German Renewable Energy Legislation.

With the purpose of protecting the environment, managing global warming and securing a reliable energy supply, the German Government and the German Bundestag committed themselves to at least doubling the percentage share of renewable energy in total energy supply by the year 2010.

This objective is related to the envisaged commitment on the part of Germany to reduce greenhouse gas emissions by 21 percent by the year 2010 in the framework of the Kyoto Protocol within the Climate Convention of the United Nations; this objective is linked to the German Government's objective to reduce carbon dioxide emissions by 25 percent by the year 2005, relative to 1990.

In order to attain this objective, it became necessary to mobilise renewable energy sources. Traditional hydropower from large dams today accounts for a significant share of renewable energy sources. However, the utilisation potential of hydropower is almost exhausted. For this reason, it is necessary to generate additional electricity from wind energy, solar energy, biomass, and small-scale hydro in order to attain the objective set for Europe as a whole by the year 2010. To this end, the currently used potential of these energy sources will have to grow fivefold.

In view of growing meteorological evidence of global warming and the increase of natural disasters world-wide, prompt action by the legislators is considered indispensable in the interest of protecting the environment and managing global warming.

Consequences of Various Legislation Models.

Three countries with price regulations, Germany, Spain, and Denmark represent over 75% of wind energy generation in Europe. Great Britain in comparison has high wind power resources, however, with quantity tender regulations as with a number of other European countries. These countries represent a modest share of wind power in Europe.

The compensation rates of the earlier German law had not been sufficient to stimulate a large-scale market introduction of electricity generated from other sources, especially photovoltaic and biomass. For this reason, the compensation rates were modified in the new Renewable Energy Sources Act, in order to promote large-scale generation of electricity from all kinds of renewable energy sources.

The new law was decided in the German Bundestag 25th February 2000 and implemented by 1st April 2000. Sufficient data were available so that a comparison was made between minimum price regions and quantity regulations. The tendering procedure used in countries with quantity regulations seems to lead to a bureaucratic and costly process, both for the operators and for the authorities. However, the main problem is that each tendering round introduces maximum limits, i.e. quotas for the various technologies.

Table 40:   Wind Energy in Europe. Comparison of price regulations with quantity regulations

	Country	Cumulated end of 2002 (MW)	Installed in 2002 [MW]	Installed capacity per area (kW/km²)	Installed capacity per capita (Watt/capita)
Countries with price regulations (Feed-in law)	Germany	12001	3247	33,6	146
	Denmark	2880	346	66,8	539
	Spain	4830	1493	9,6	122
	Sum	**19711**	**5086**		
Countries with quantity regulations (Call for tenders)	UK	552	78	2,2	9
	Ireland	137	12	1,9	36
	France	148	32	0,3	3
	Sum	**837**	**124**		

Source: Folkecenter for Renewable Energy and New Energy 2/2003

As a result, only part of the technical and economic potentials is exploited. This impedes natural market developments. Also, the pressure on costs due to competition between tenderers means that only large-scale projects are carried out and in areas with very favourable wind conditions. Small private operators or local communities, leading the implementation in Germany or Denmark, are practically excluded from the tendering process, because they cannot keep up financially with the big central planning and developing companies, who are often backed by banks and energy suppliers with strong financial resources.

In UK implementation principles are still preferred and may be successfully applied for multi-megawatt projects in un-populated regions and off-shore, but this will cause obstacles and objection from citizens in high-density areas including most of Europe, Japan, and parts of USA. In UK, this has led to significant local acceptance problems for wind power in contrast to the occasional but harmless local frictions in Germany or Denmark. The local population in UK is not included in the project planning and implementation and is not directly benefiting economically. So they often use their civil rights to protest against plans of wind energy to avoid the visual impact on their neighbourhood.

Finally, tenderers often submit unrealistically low tenders owing to the strong competition. In most cases the projects in UK are not carried out. Even if in UK wind resources are substantially much better than in Germany, less than 600 MW was installed in UK by end of 2002 compared to 12,000 MW in Germany. Both countries have access to the same wind energy technology from the leading manufacturers so the difference in implementation of a factor of 20 can only be explained by choosing un-appropriate policies and strategies that cause barriers to wind energy instead of stimulation for market development and a home market for wind mill manufacturers.

The policies and principles of the UK seem, irrespective of their evident defects in achieving official targets of implementation, to have been the model for the initial draft renewable energy directive of the European Commission in 1998. It was, however, modified during the decision process to incorporate feed-in tariff principles before ultimately being approved by the Council of Ministers in 2001. The crucial argument against the quantity-rated tendering model, however, is the official stipulation of a specific upper limit for renewable energies due to uncertainty whether it will be extended by political decision. A market where further development depends on government decisions, while at the same time no defined limits are set to the inputs of nuclear or fossil energy, adds up to discrimination against the new, clean but capital intensive energy solutions.

Furthermore it is not realistic to expect a government to increase the quota without extensive public debates and objection from vested interests. Each redefinition of the quota collides with conventional power producers' interest in preserving their share of the electricity market by maintaining status quo. Traditionally, these producers have strong influence in all countries on the political process.

Limitations and Exemptions of the New Legislation
The new German renewable energy law shall not apply to electricity:

1. Produced by hydro-electric power plants and installations fuelled by gas from landfills or sewage treatment plants with an installed electrical capacity of over 5 MW, or by installations in which electricity is generated from biomass, with an electrical capacity of over 20MW.

2. Produced by installations of which over 25% is owned by the Federal Republic of Germany or one of Germany's federal states.

3. Produced by installations for the generation of electricity from solar energy, with an installed electrical capacity of over five megawatts. In the case of installations for electricity from solar energy that are not part of structures for purposes other than the generation of electricity, the upper capacity limit shall be 100 KW.

Obligation to Purchase and Pay Compensation.

Grid operators shall be obliged to connect to their grids installations to purchase electricity from these installations as a priority, and to compensate the suppliers of this electricity. This obligation shall apply to the grid operator, whose grid is closest to the location of the electricity generation installation, providing that the grid is technically suitable to feed-in this electricity.

The upstream transmission grid operator shall be obliged to purchase, and pay compensation for the amount of energy purchased by the grid operator. If there is no domestic transmission grid in the area serviced by the grid operator entitled to sell electricity, the next closest domestic transmission grid operator shall be obliged to purchase and pay compensation for this electricity. The minimum compensation amounts shall be payable for newly commissioned installations for a period of 20 years after the year of commissioning, except for installations which generate electricity from hydropower.

Table 41: Global Wind Energy Statistics. Status per December 2002.

Country	Wind energy end of 2002 [MW]	Wind energy end of 2001 [MW]	Wind energy end of 2000 [MW]	Wind energy end of 1999 [MW]	Wind energy 2002* [MW]	Rate of growth* [%]	Population 2000 [millions]	Installed capacity per capita [Watt/capita]	Area [km²]	Installed capacity per area [kW/km²]	Gross National Product GNP 2000 [bn. $]
Germany	12,001	8,754	6,095	4,443	3,247	37.1	82.15	146.09	357,020	33.6144	1,872.99	6.41
Spain	4,830	3,337	2,502	1,476	1,493	44.7	39.45	122.43	504,782	9.5685	558.56	8.65
USA	4,685	4,275	2,581	2,551	410	9.6	281.55	16.64	9,809,155	0.4776	9,837.41	0.48
Denmark	2,880	2,383	2,306	1,700	497	20.9	5.34	539.33	43,096	66.8275	162.34	17.74
India	1,702	1,507	1,267	1,035	195	12.9	1,015.92	1.68	3,287,263	0.5178	456.99	3.72
Italy	785	682	418	283	103	15.1	57.68	13.60	301,336	2.6041	1,073.96	0.73
The Netherlands	686	484	443	411	202	41.7	15.92	43.09	41,526	16.5198	364.77	1.88
United Kingdom	552	474	409	347	78	16.5	59.74	9.24	242,910	2.2724	1,414.56	0.39
China*	468	400	352	262	68	17.0	1,261.10	0.37	9,572,419	0.0489	1,079.95	0.43
Japan*	415	275	142	68	140	50.9	126.77	3.27	377,837	1.0984	4,841.58	0.09
Sweden	328	293	241	215	35	11.9	8.87	36.98	449,964	0.7289	227.32	1.44
Greece	276	272	247	158	4	1.5	10.56	26.14	131,957	2.0916	112.65	2.45
Canada	221	207	128	124	14	6.8	30.74	7.19	9,984,670	0.0221	687.88	0.32
Portugal	194	131	83	67	63	48.1	10.01	19.38	92,345	2.1008	105.06	1.85
France	148	95	68	25	53	55.8	58.85	2.51	543,965	0.2721	1,294.25	0.11
Austria	139	94	77	42	45	47.9	8.09	17.18	83,858	1.6576	189.03	0.74
Ireland	137	125	119	73	12	9.6	3.79	36.15	70,273	1.9495	93.87	1.46
Australia	104	72	30	9	32	44.4	19.20	5.42	7,692,300	0.0135	390.11	0.27
Norway*	97	17	13	13	80	462.3	4.49	21.60	323,759	0.2996	161.77	0.60
Costa Rica*	71	71	51	51	0	0.0	3.64	19.53	51,060	1.3905	15.85	4.48
Egypt*	69	69	69	36	0	0.0	63.82	1.08	1,002,000	0.0689	98.73	0.70
Morocco*	54	54	54	14	0	0.0	28.71	1.88	458,730	0.1177	33.35	1.62
Belgium	44	32	13	9	12	37.5	10.25	4.29	32,545	1.3520	226.65	0.19
Ukraine*	44	41	5	5	3	7.3	49.60	0.89	603,700	0.0729	31.79	1.38
Finland	41	39	38	38	2	5.1	5.18	7.92	338,144	0.1213	121.47	0.34
New Zealand*	35	35	35	35	0	0.0	3.83	9.14	270,534	0.1294	49.90	0.70
Poland*	27	22	5	5	5	22.7	38.65	0.70	312,685	0.0863	157.74	0.17
Argentina*	27	27	16	15	0	1.7	37.03	0.73	2,780,400	0.0097	284.96	0.09
Brazil*	25	24	22	19	1	2.5	170.12	0.15	8,547,404	0.0029	595.46	0.04
Latvia*	24	2	0	0	22	1,100.0	2.42	9.93	64,589	0.3716	7.15	3.36
Turkey*	19	19	19	9	0	0.0	65.31	0.29	779,452	0.0244	199.94	0.10
Luxembourg	16	15	15	10	1	6.7	0.44	36.36	2,586	6.1872	18.89	0.85
Iran*	11	11	11	11	0	0.0	64.02	0.17	1,648,000	0.0067	104.90	0.10
Tunisia*	11	11	11	0	0	0.0	9.58	1.15	163,610	0.0672	19.46	0.57
Sum	**31,166**	**24,349**	**17,885**	**13,559**	**6,817**	**28.0**					

(right margin label: Installed capacity per GNP [MW/bn. $])

Source: New Energy 2/2003

Compensation for Electricity from Biomass.

1. At least 10.23 € Cent per kWh for installations with an installed electrical capacity up to 500 KW.

2. At least 9.21 € Cent per kWh for installations with an installed electrical capacity up to 5 MW.

3. At least 8.70 € Cent per kWh for installations with an installed effective electrical capacity over 5 MW.

The use of biomass, including biogas, for electricity generation represents an insufficiently used potential. At the same time, biomass provides additional perspectives for domestic agriculture and forestry.

The compensation rates have been increased substantially above the rates laid down in the 1991 law in order to enable operators of biomass installations to operate their installations cost-effectively, thereby initiating a dynamic development. Compensation rates differ in accordance with the electrical capacity giving a premium to smaller decentralised installations.

Compensation for Electricity from Wind Energy.

The new law introduces a differentiated set of tariffs for wind power.

According to the intentions of the law, wind power in Germany was projected to deliver 22 TWH in 2007, which is 5% of the total national power consumption. This target was achieved in 2002/2003 illustrating the dynamics of the law.

1. The electricity generated from wind energy shall be at least 9.10 € Cent per kWh for a period of five years from the date of commissioning and thereafter gradually reduced following specific guidelines.

2. For off-shore wind farms commissioned before 2007, the period with the high compensation 9.10 € Cent per kWh is increased to nine years.

3. As of 1 January 2002, the minimum compensation amounts shall be reduced by 1.5 % annually for new installations commissioned after this date.

The previous provisions applying to wind energy did not consider the differences between various sites. The future rates will vary as a function of site profitability. The new provisions lead to the following results: at very good sites, compensation rates will be reduced to 6.90 € Cent per kWh for a 20 years period; at sites with reference average wind conditions, the rates will be at 8.39 € Cent per kWh, and at inland sites, the rates will be moderately changed to 8.85 € Cent per kWh during the 20 years of operation.

Figure 178: Two out of 20 off-shore windmills, each 2 MW, at Middelgrunden between Sweden and Denmark. Photo by V. Kantor.

The intention of the law is to avoid payment of compensation rates that are higher than what is required for a cost-effective operation and to create an incentive for installing windmills at inland sites. The higher initial compensation rate will facilitate the financing of windmills that were previously being questioned by credit institutions. The electricity production costs of off-shore wind projects are expected to decrease substantially in the future. Initially, however, the investment cost is higher than the cost of on-shore installations due to the lack of experience, higher expenses for new technology, complicated foundation systems and the lack of economies of scale. The provisions apply to wind turbines located at least three miles seaward from the baselines.

Compensation for Electricity from Solar Power.

In the long term, the use of solar electricity holds the greatest potential for providing energy supply without adverse impacts on the climate. This energy source requires sophisticated technology and high industrial importance in the future. There are relatively high rates because solar-generating installations were still not yet produced in sufficient quantities in 2000.

1. The compensation for electricity solar energy shall be at least 50.62 € Cent per kWh. As of 1 January 2002, the minimum price paid shall be reduced by 5% annually for new electricity.

2. The obligation to pay compensation shall not apply to photovoltaic installations commissioned after 31 December of a year and exceeding 350 MW.

3. The compensation applies only to installations integrated in buildings and only up to 100 KW peak capacity providing optimal consequences for industrial development, job creation and a new building culture.

As soon as the new law has created sufficient demand, larger production volumes can be expected to lead to a substantial reduction in manufacturing cost, and hence, in electricity production cost, so that the compensation rates can be allowed to decrease accordingly. In combination with the "100,000 Roofs Programme", the compensation payments for the first time will make electricity generation from solar energy an attractive option for private investors; however, in many cases, the compensation specified does not permit a profitable operation of such installations at all times.

Figure 179: Solar architecture and photovoltaics at Reichstag, German Parliament, Berlin, by Sir Norman Foster (left); excursion ship by Kopf at Alster lake, Hamburg, Germany, driven by roof mounted solar panels (right).

The Innovative Aspects of the Law.

The compensation defined in the new Renewable Energy Sources Act is based on the systematic approach introduced in the Electricity Feed Act and following the recommendations presented by the European Commission in its White Paper, "Energy for the Future: Renewable Sources of Energy" as well as the relevant resolutions adopted by the European Parliament. The compensation rates specified in the Renewable Energy Sources Act have been determined by means of scientific studies, so that the rates identified should make it possible for an installation to be operated cost-effectively, based on the use of state-of-the-art technology and depending on the renewable energy sources naturally available in a given geographical environment.

In some cases, the cost of the production of renewable energy sources is higher than the production cost of conventional energy sources because the overwhelming share of the external costs associated with electricity from conventional energy sources is not reflected in the price; instead, these costs are borne by the general public and by future generations Furthermore costs for military security of oil exploration and supply is not included. In addition, conventional energy sources benefit from substantial governmental subsidies that keep their price artificially low. Globally this amounts to 300 billion EUR per year. Another reason for the higher costs is the structural discrimination of new technologies. Their lower market share does not allow economies of scale to become effective.

For this reason, the purpose of the new law was not only to protect the operation of existing installations but also to stimulate a dynamic development in all fields of electricity generation from renewable energy sources.

In combination with measures aimed at internalising external costs, the purpose of the pricing regime is to bring renewable energy sources closer to conventional energy sources in terms of their competitiveness. In order to continue to facilitate major improvements in technological efficiency, the compensation rates specified in the Renewable Energy Sources Act vary, depending on the energy sources, the sites and the installation sizes involved. Furthermore, they will decline over time and will remain in effect for a limited period of time. The fact that the rates will be reviewed every two years guarantees that they will be updated continuously and at short intervals to reflect market and cost trends. Compensation paid under this law is not state aid from a terminological perspective because producers of renewable energy are not granted special benefits; the law compensates disadvantages that such operators have in comparison with conventional electricity producers.

Most of the social and ecological external follow-up costs associated with conventional electricity generation are currently not borne by the operators of such installations but by the general public, the taxpayers and future generations. The Renewable Energy Sources Act reduces the competitive advantage that conventional power producers have enjoyed compared to renewable energy producers which causes limited or no external costs.

Importance of political priorities and further growth for wind energy.

The successful wind energy development worldwide clearly reflects the political priority that the leading countries are giving to wind energy by creating favourable frameworks. In several countries, wind energy is already contributing substantially to the electricity supply, reaching in some regions shares of 50 % and more. Regarding the international efforts in climate protection and the limitation of fossil and nuclear resources, it is obvious that the steady growth of wind energy utilisation will be continued in the future. More and more governments see the need to utilise local renewable resources instead of relying on imported fossil and nuclear resources with their high political and economic risks.

Based on a WWEA member survey conducted in February 2004, 100,000 MW of wind energy capacity is expected to be overstepped in the year 2008. Further substantial growth can in the next years be expected in the five currently leading countries, as well as in further European and in several Asian countries. Especially the emerging countries with their high need for additional energy sources to cover their economic growth like China, India and Brazil are now more and more focussing on renewable energies. In the industrialised countries, the driving forces are mainly: environmental and climate protection, security of the energy supply and economic advantages of a decentralised energy supply. For developing countries, very interesting perspectives for electrification of un-served areas lie in the utilisation of decentralised, small-scaled off-grid applications.

Appendices

1 Glossary

Regional Definitions

In this Statement the world is divided into two main groups:
- the market oriented economies (both developed and developing), and
- the economies in transition (which include both industrialised and developing economies).

The developed market economies are, with the exception of Turkey, those of the OECD at the time when ETW was published in 1993. They include North America (US and Canada), Iceland, Western Europe (including the former Eastern part of Germany but excluding Turkey) and the Pacific countries (Japan, Australia and New Zealand). It should be noted that the OECD now includes three economies in transition (the Czech Republic, Hungary and Poland) and two developing countries (the Republic of Korea and Mexico).

The developing market economies include Latin America (which includes Mexico and the Caribbean), Africa, Asia (including Turkey and the Middle East but excluding the economies in transition of Asia).

The developed economies in transition are the countries of Central and Eastern Europe (CEE), the Commonwealth of Independent States (CIS), and other republics of the Former Soviet Union. We recognise that some of these dividing lines, for instance in Central Asia, between industrialised and developing economies are extremely fine.

The developing economies in transition include China, Cambodia, DPR of Korea, DPR of Laos, Vietnam and Mongolia.
WEC now uses the following regional divisions:

- Africa, which includes North Africa from Algeria to Egypt.
- Asia, which includes Asia Pacific, Middle East and West Asia (including Turkey), and South Asia.
- Europe, which includes Central and Eastern Europe, the CIS, and Western Europe.
- Latin America and the Caribbean (including Mexico).
- North America (US and Canada).

Accuracy
The degree to which the mean of a sample approaches the true mean of the population; lack of bias.

Activity
A practice or ensemble of practices that take place on a delineated area over a given period of time.

Adaptation
Adjusting natural or human systems to cope with actual or expected climate change and its impacts.

Additives
Chemicals added to fuel in very small quantities to improve and maintain fuel quality. Detergents and corrosion inhibitors are examples of gasoline additives.

Additionality
A project is *additional* if it would not have happened, but for the incentive provided by the credit trading program (e.g. CDM or JI). The Kyoto Protocol's specifies that only projects that provide emission reductions that are *additional* to any that would occur in the absence of the project activity shall be awarded certified emission reductions (CERs) in the case of CDM projects or emission reduction units (ERUs) in the case of JI projects. This is often referred to as "environmental additionality".

Advanced Technology Vehicle (ATV)
A vehicle that combines new engine/power/drive train systems to significantly improve fuel economy. This includes hybrid power systems and fuel cells, as well as some specialized electric vehicles.

Agriculture, Small Scale
Ways to improve crop yields in small areas; permaculture and proven, new techniques.

Agriculture, Larger Scale
New farming methods for high altitude and specific growing conditions.

Air Toxics
Toxic air pollutants, including benzene, formaldehyde, acetaldehyde, 1-3 butadiene and polycyclic organic matter (POM). Benzene is a constituent of motor vehicle exhaust, evaporative and refuelling emissions. The other compounds are exhaust pollutants.

Alcohols
Organic compounds that are distinguished from hydrocarbons by the inclusion of a hydroxyl group. The two simplest alcohols are methanol and ethanol.

Aldehydes
A class of organic compounds derived by removing the hydrogen atoms from an alcohol. Aldehydes can be produced from the oxidation of an alcohol.

Alternative Fuel
Methanol, denatured ethanol, and other alcohols; mixtures containing 85% or more by volume of methanol, denatured ethanol, and other alcohols with gasoline or other fuels; natural gas; liquefied petroleum gas; hydrogen; coal-derived liquid fuels; non-alcohol fuels (such as biodiesel) derived from biological material; and electricity.

Alternative energy
Energy derived from non-fossil fuel sources.

American Society for Testing and Materials (ASTM)
A non-profit organization that provides a management system to develop published technical information. ASTM standards, test methods, specifications and procedures are recognized as definitive guidelines for motor fuel quality as well as a broad range of other products and procedures.

Ancillary effects
Side effects of policies to reduce net greenhouse gas emissions, such as reductions in air pollutants associated with fossil fuels or socio-economic impacts on employment or agricultural efficiency.

Anhydrous
Describes a compound that does not contain any water. Ethanol produced for fuel use is often referred to as anhydrous ethanol, as it has had almost all water removed.

Animal Husbandry
Taking care of animals, improving strains, basic veterinary skills.

Anthropogenic
Man made.

Anthropogenic emissions
Greenhouse gas emissions associated with human activities such as burning fossil fuels or cutting down trees.

Annex I Parties
Originally thirty-five countries, comprising industrialised economies (including Turkey which has since sought to withdraw voluntarily from the Climate Convention) and eleven transitional economies plus the then European Economic Community as a regional organisation, that are Parties to the UN Framework Convention on Climate Change and so listed in Annex I to the Convention. Since 1992 the withdrawal of Turkey and the creation of separate Czech and Slovakian Republics, Croatia and Slovenia, have modified the number of countries.

Annex B Parties
Thirty-nine countries listed in Annex B to the Kyoto Protocol which indicated agreement at the Third Conference of the Parties to the UN Climate Convention to contemplate legally binding quantified emission limitation and reduction commitments. The Kyoto Protocol has not yet been ratified.

Aquaculture
Renewable energy powered fish farming for nutrition, health, help in ending hunger and micro-enterprise.

Aromatics
Hydrocarbons based on the ringed six-carbon benzene series or related organic groups. Benzene, Toluene and Xylene are the principal aromatics, commonly referred to as the BTX group. They represent one of the heaviest fractions in gasoline.

Array
Often several modules are wired together in an array to increase the total available power output and to match the operating voltage of the "load" equipment.

Balance of Payments
The dollar amount difference between a country's exports and imports. In the United States, large oil imports are one of the main causes of the negative balance of payments with the rest of the world.

Barrier
Any obstacle to the diffusion of cost-effective mitigation technologies or practices, whether institutional, social, economic, political, cultural or technological.

Baseline
The greenhouse gas emissions level that would occur in the absence of climate change interventions; used as a basis for analysing the effectiveness of mitigation policies.

Baseline scenario
A baseline scenario is a presumed counterfactual alternative to the proposed project. In other words, it is an interpretation of "what would have happened otherwise". Several plausible baseline scenarios can be evaluated for a given project. The project itself can and should typically be considered as one of these baseline scenarios, since the possibility it would have been implemented in the absence of carbon credits must be examined to determine whether it is additional.

Baseline emission rate
A baseline emission rate is the parameter, expressed in t CO_2/MWh for electricity projects, which is used to calculate the number of emission credits (i.e. CERs or ERUs) a project can generate. The baseline emissions rate can be based on standardised (multi-project) methodologies, or correspond directly to a project specific baseline scenario.

Baseline scenario analysis
This commonly-used baseline methodology involves a process of elaborating, then culling through, a series of plausible baseline scenarios, often involving investment or barrier analysis. This analysis is useful to demonstrate the *additionality* of a project.

Base load
That part of total energy demand that does not vary over a given period of time.

Benzene
A six-carbon aromatic; common gasoline component identified as being toxic. Benzene is a known carcinogen.

Bias
Systematic over- or under-estimation of a quantity.

Biochemical Conversion
The use of enzymes and catalysts to change biological substances chemically to produce energy products. For example, the digestion of organic wastes or sewage by micro organisms to produce methane is a biochemical process.

Biodiesel
A biodegradable transportation fuel for use in diesel engines that is produced through transesterification of organically derived oils or fats. Biodiesel is used as a substitute of diesel fuel.

Biofuels
Fuel produced from dry organic matter or combustible oils from plants, such as alcohol from fermented sugar, black liquor from the paper manufacturing process, wood, and soybean oil.

Biogas
Methane rich gas containing 30 – 45% CO2 and various trace elements especially H2S. Biogas is produced from organic wastes and residues from agriculture and households by anaerobic fermentation in digesters of sizes from 2 to 3000 cubicmeters and temperatures
between 20 and 50 C.

Bio-Generator Power System
Village-scale power in remote locations without a fuel source, much less expensive than solar or wind, uses available agricultural waste.

Biological options
There are three: conserving an existing carbon pool, and thereby preventing emissions into the atmosphere; sequestrating more CO_2 from the atmosphere by increasing the size of existing carbon pools; and substituting biological products for fossil fuels or for energy-intensive products, thereby reducing CO_2 emissions.

Biomass
The total mass of living organisms in a given area or volume; biomass can be used as a sustainable source of fuel with low or zero net emissions.

Biosphere
That component of the Earth system that contains life in its various forms, which includes its living organisms and derived organic matter (e.g., litter, detritus, soil).

British Thermal Unit (Btu)
A standard unit for measuring heat energy. One Btu represents the amount of heat required to raise one pound of water one degree Fahrenheit (at sea level).

Briquetting
Making a more efficient, smokeless fuel source with local materials and without wood, keeping dung for gardens and pastures instead of going up in smoke.

BTX
Industry term referring to the group of aromatic hydrocarbons benzene, toluene and xylene (see aromatics).

Build margin
The build margin refers to new sources of electric capacity expected to be built or otherwise added to the system, and affected by a new project-based activity.

Butane
A gas, easily liquefied, recovered from natural gas. Used as a low-volatility component of motor gasoline, processed further for a high-octane gasoline component, used in LPG for domestic and industrial applications and used as a raw material for petrochemical synthesis.

Butyl Alcohol
Alcohol derived from butane that is used in organic synthesis and as a solvent.

Carbon Dioxide (CO_2)
The gas formed in the ordinary combustion of carbon, given out in the breathing of animals, burning of fossil fuels, etc. Human sources are very small in relation to the natural cycle.

Carbon Monoxide (CO)
A colourless, odourless gas produced by the incomplete combustion of fuels with a limited oxygen supply, as in automobile engines.

Carbon Flux
Transfer of carbon from one carbon pool to another in units of measurement of mass per unit area and time (e.g., $t\ C\ ha^{-1}\ y^{-1}$).

Carbon Pool
A reservoir. A system, which has the capacity to accumulate or release carbon. Examples of carbon pools are forest biomass, wood products, soils, and atmosphere. The units are mass (e.g., $t\ C$).

Carbon Sequestration
The absorption and storage of CO_2 from the atmosphere by the roots and leaves of plants; the carbon builds up as organic matter in the soil.

Carbon Stock
An absolute quantity of carbon held within a pool at a specified time.

Carcinogens
Chemicals and other substances known to cause cancer.

Catalyst
A substance whose presence changes the rate of chemical reaction without itself undergoing permanent change in its composition. Catalysts may be accelerators or retarders. Most inorganic catalysts are powdered metals and metal oxides, chiefly used in the petroleum, vehicle and heavy chemical industries.

Cetane
Ignition performance rating of diesel fuel. Diesel equivalent to gasoline octane.

Closed-Loop Carburetion
System in which the fuel/air ratio in the engine is carefully controlled to optimise emissions performance. A closed-loop system uses a fuel metering correction signal to optimise fuel metering.

Co-solvents
Heavier molecular weight alcohols used with methanol to improve water tolerance and reduce other negative characteristics of gasoline/alcohol blends.

Combined margin baselines
A combined margin baseline reflects both operating and build margin effects.

Communications
Bridging the "digital divide"; satellite, long range phone, and internet communication systems.

Commercial Energy
Energy supplied on commercial terms. Distinguished from non-commercial energy comprising fuel wood, agricultural wastes and animal dung collected usually by the user.

Community Organization
Mobilizing community involvement, organizing a task-force, and developing a five year village development plan

Compressed Natural Gas (CNG)
Natural gas that has been compressed under high pressures, typically 2.000 ue metricto 3.600 psi, held in a container. The gas expands when used as a fuel.

Compression Ignition or Diesel Engine
The form of ignition that initiates combustion in a diesel engine. The rapid compression of air within the cylinders generates the heat required to ignite the fuel as it is injected.

Converted or Conversion Vehicle
A vehicle originally designed to operate on gasoline or diesel that has been modified or altered to run on an alternative fuel.

Cooking Oil Production
Hand and solar-powered presses to make everything from cooking oil to bio-fuels that can power vehicles.

Cooling (Refrigeration) Systems

How to make and use "pot-within-a-pot" earthenware cooling systems with locally available materials (clay pots, sand and cloth). In dry climates, using these super-inexpensive systems, vegetables stay fresh for weeks instead of just a few days.

Corrosion Inhibitors

Additives used to inhibit corrosion in the fuel system (e.g., rust).

Cryogenic Storage

Extreme low-temperature storage.

Dedicated Natural Gas Vehicle

A vehicle that operates only on natural gas. Such a vehicle is incapable of running on any other fuel.

Dedicated Vehicle

A vehicle that operates solely on one fuel. Generally, dedicated vehicles have superior emissions and performance results because their design has been optimised for operation on a single fuel.

Denatured Alcohol

Ethanol that contains a small amount of a toxic substance, such as methanol or gasoline, which cannot be removed easily by chemical or physical means. Alcohols intended for industrial use must be denatured to avoid federal alcoholic beverage tax.

Detergent

Additives used to inhibit deposit formation in the fuel and intake systems in automobiles.

Direct emissions

Direct emissions are a direct consequence of project activity, either on-site (e.g. via fuel combustion at the project site) or off-site (e.g. from grid electricity or district heat, and other upstream and downstream life cycle impacts).

Distillation Curve

The percentages of gasoline that evaporate at various temperatures. The distillation curve is an important indicator for fuel standards such as volatility (vaporization).

Distributed generation

This term is used for a generating plant serving a customer on-site, or providing support to a distribution network, and connected to the grid at distribution level voltages. The technologies generally include engines, small (including micro) turbines, fuel cells, and photovoltaics.

Dual-Fuel Vehicle
Vehicle designed to operate on a combination of an alternative fuel and a conventional fuel. This includes a) vehicles using a mixture of gasoline or diesel and an alternative fuel in one fuel tank, commonly called flexible-fuelled vehicles; and b) vehicles capable of operating either on an alternative fuel, a conventional fuel or both, simultaneously using two fuel systems commonly called bi-fuel vehicles.

Dynamometer
An instrument for measuring mechanical force, or an apparatus for measuring mechanical power (as of an engine).

E10 (Gasohol)
Ethanol mixture containing 93% ethanol, 5% methanol and 2%kerosene, by volume.

E85
Ethanol/gasoline mixture containing 85% denatured ethanol and 15% gasoline, by volume.

E93
Ethanol mixture containing 93% ethanol, 5% methanol and 2% kerosene, by volume.

E95
Ethanol/gasoline mixture containing 95% denatured ethanol and 5% gasoline, by volume.

Eco-Tourism
Ways to develop an income by serving tourists; clean water, showers, clean beds, internet service.

Education
Training the trainers; educational methodologies for assuring the best participation and success of project implementation.

Electric Vehicle
A vehicle powered by electricity, generally provided by batteries. EVs qualify in the zero emission vehicle (ZEV) category for emissions.

Electricity
Electric current used as a power source. Electricity can be generated from a variety of feedstocks including oil, coal, nuclear, hydro, natural gas, wind, and solar. In electric vehicles, onboard rechargeable batteries power an electric motor.

El Niño
A warm ocean surge in the East Pacific off the coast of Peru which has occurred every few years for thousands of years, and which regularly causes major climatic disruption over a wide area. Often referred to as an ENSO (El Niño Southern Oscillation) event. The reverse phase of El Niño is referred to as La Niña.

Emissions tax
A levy imposed by a government on each unit of CO_2 equivalent emissions from a source subject to the tax; can be imposed as a carbon tax to reduce carbon dioxide emissions from fossil fuels.

Emissions trading
A market-based approach to achieving environmental objectives that allows countries or companies that reduce greenhouse gas emissions below their target to sell their excess emissions credits or allowances to those that find it more difficult or expensive to meet their own targets.

Energy Intensity
The proportion of energy used to Gross Domestic Product at constant prices.

Ester
An organic compound formed by reacting an acid with an alcohol, always resulting in the elimination of water.

Ethane (C_2H_6)
A colourless hydrocarbon gas of slight odour having a gross heating value of 1,773 BT. It is a normal constituent of natural gas.

Ethanol (also known as Ethyl Alcohol, Grain Alcohol, CH 3 CH 2 OH)
Can be produced chemically from ethylene or biologically from the fermentation of various sugars from carbohydrates found in agricultural crops and cellulosic residues from crops or wood. Used in the United States as a gasoline octane enhancer and oxygenate, it increases octane 2.5 to 3.0 numbers at 10% concentration. Ethanol also can be used in higher concentration in alternative-fuel vehicles optimised for its use.

Ether
A class of organic compounds containing an oxygen atom linked to two organic groups.

Etherification
Oxygenation of an olefin by methanol or ethanol. For example, MTBE is formed from the chemical reaction of isobutylene and methanol.

Ethyl Alcohol
See Ethanol.

Ethyl Ester
A fatty ester formed when organically derived oils are combined with ethanol in the presence of a catalyst. After water washing, vacuum drying and filtration, the resulting ethyl ester has characteristics similar to petroleum-based diesel motor fuels.

Ethyl Tertiary Butyl Ether (ETBE)
A fuel oxygenate used as a gasoline additive to increase octane and reduce engine knock.

Evaporative Emissions
Hydrocarbon vapours that escape from a fuel storage tank or a vehicle fuel tank or vehicle fuel system.

Flexible-Fuel Vehicle
Vehicles with a common fuel tank designed to run on varying blends of unleaded gasoline with either ethanol or methanol.

Fluidised Beds
Beds of burning fuel and non-combustible particles kept in suspension by upward flow of combustion air through the bed. Limestone or coal ashes are widely used non-combustible materials.

Flux
See "Carbon Flux".

Food Storage
Solar crop drying methods, Butter/cheese storage, meat preservation; coffee, nuts, grain.

Forest Stand
A community of trees, including aboveground and belowground biomass and soils, sufficiently uniform in species composition, age, arrangement, and condition to be managed as a unit.

Fossil fuels
Carbon-based fuels from fossil carbon deposits, including coal, oil, and natural gas.

Fuel Cell
An electrochemical engine with no moving parts that converts the chemical energy of a fuel, such as hydrogen, and an oxidant, such as oxygen, directly to electricity. The principal components of a fuel cell are catalytically activated electrodes for the fuel (anode) and the oxidant (cathode) and an electrolyte to conduct ions between the two electrodes.

Gasifiers
Tank for anaerobic fermentation of biomass residues from sugar cane, pulp and paper, etc, to produce biogas.

Gasoline Gallon Equivalent (gge)
A unit for measuring alternative fuels so that they can be compared with gasoline on an energy equivalent basis. This is required because the different fuels have different energy densities.

Geothermal
Natural heat extracted from the earth's crust using its vertical thermal gradient, most readily available where there is a discontinuity in the earth's crust (e.g. where there is separation or erosion of tectonic plates).

Global Warming
The theoretical escalation of global temperatures caused by the increase of greenhouse gas emissions in the lower atmosphere.

Greenhouse Effect
A warming of the earth and its atmosphere as a result of the thermal trapping of incoming solar radiation by CO_2, water vapour, methane, nitrous oxide, chlorofluorocarbons, and other gases, both natural and man-made.

Greenhouse Gases
Gases which, when concentrated in the atmosphere, prevent solar radiation trapped by the Earth and re-emitted from its surface from escaping. The result is *ceteris paribus*, a rise in the Earth's near surface temperature. The phenomenon was first described by Fourier in 1827, and first termed the greenhouse effect by Arrhenius in 1896. Carbon dioxide is the largest in volume of the greenhouse gases. The others are halocarbons, methane, nitrous oxide, hydro fluorocarbons, perfluorocarbons, and sulphur hexafluoride.

Gross Vehicle Weight (gvw)
Maximum weight of a vehicle, including payload.

Halocarbons
A family of chlorofluorocarbons (CFCs) mostly of industrial origin – CH_3Cl is the main exception. Includes aerosol propellants (CFCs 11, 12, 114); refrigerants (CFCs 12, 114 and HCFC-22); foam-blowing agents (CFCs 11 and 12); solvents (CFC-113, CH_3CCl_3 and CCl_4); and fire retardants (halons 1211 and 1301). HCFC = hydrofluorocarbon.

Heterotrophic Respiration
The release of carbon dioxide from decomposition of organic matter.

Hybrid Electric Vehicle
A vehicle powered by two or more energy sources, one of which is electricity. HEVs may combine the engine and fuel of a conventional vehicle with the batteries and electric motor of an electric vehicle in a single drive train.

Indirect emissions
Indirect emissions occur when market and individual response to a project activity leads to increased or decreased emissions. Indirect emissions can be either on-site (e.g. rebound effects such as increased heating that may result from of an insulation program) or off-site (e.g.* project effects that are often referred to as leakage, either negative or positive, such as economy-wide response to price changes or increased penetration of low carbon technologies outside the project site induced by the project activity).

Internet Café
Ways to create micro-enterprise services from a satellite Internet connection.

Investment additionality
Investment additionality seeks to compare the financial return of a project and its alternative (baseline scenario) to determine additionality of a CDM or JI project and/or the most likely baseline.

Land Cover
The observed physical and biological cover of the Earth's land as vegetation or man-made features.

Land Use
The total of arrangements, activities, and inputs undertaken in a certain land cover type (a set of human actions). The social and economic purposes for which land is managed (e.g., grazing, timber extraction, conservation).

Lead
see Tetraethyl Lead.

Leakage
Occurs when emissions reductions in developed countries are partly off-set by increases above baseline levels in developing countries, due to relocation of energy-intensive production, increased consumption of fossil fuels when decreased developed country demand lowers international oil prices, or changes in incomes and thus in energy demand because of better terms of trade, or when sink activities such as tree planting on one parcel of land encourage emitting activities elsewhere.

LED / Hand-Generator Manufacturing
The least expensive lighting systems for remote villages, a proven, small-scale manufacturing opportunity.

LNG to CNG Station
A station, supplied with LNG, that pumps and vaporizes the liquid supply to vehicles as CNG fuel, generally at the correct pressure and temperature (i.e., the temperature effect of compression is factored into the design).

LNG vehicle
A vehicle that uses LNG as its fuel.

Light-Duty Vehicle
Passenger cars and trucks with a gross vehicle weight rating of 3.500 kiligramm or less.

Liquefied Natural Gas (LNG)
Compressed natural gas that is cryogenically stored in its liquid state.

Liquefied Petroleum Gas (LPG)
A mixture of hydrocarbons found in natural gas and produced from crude oil, used principally as a feedstock for the chemical industry, home heating fuel, and motor vehicle fuel. Also known by the principal constituent propane.

Litre (L)
A metric measurement used to calculate the volume displacement of an engine. One litre is equal to 1,000 cubic centimetres or 61 cubic inches.

Lubricity
Capacity to reduce friction.

M100
100% (neat) methanol.

M85
85% methanol and 15% unleaded gasoline by volume, used as a motor fuel in FFVs.

Methane (CH_4)
A gas emitted from coal seams, natural wetlands, rice paddies, enteric fermentation (gases emitted by ruminant animals), biomass burning, anaerobic decay or organic wastes in landfill sites, gas drilling and venting, and the activities of termites.

Methanol (also known as Methyl Alcohol, Wood Alcohol, CH_3, OH)
A liquid fuel formed by catalytically combining CO with hydrogen in a 1 to 2 ratio under high temperature and pressure. Commercially, it is typically manufactured by steam reforming natural gas. Also formed in the destructive distillation of wood.

Methyl Alcohol
See Methanol.

Methyl Ester
A fatty ester formed when organically derived oils are combined with methanol in the presence of a catalyst. Methyl Ester has characteristics similar to petroleum-based diesel motor fuels.

Methyl Tertiary Butyl Ether (MTBE)
A fuel oxygenate used as an additive to gasoline to increase octane and reduce engine knock.

Mobile Source Emissions
Emissions resulting from the operations of any type of motor vehicle.

Motor Octane
The octane as tested in a single-cylinder octane test engine at more severe operating conditions. Motor Octane Number (MON) affects high-speed and part-throttle knock and performance under load, passing, climbing and other operating conditions. Motor octane is represented by the designation M in the (R+M)/2 equation and is the lower of the two numbers.

Micro-Finance
How to set up and organize a micro-lending organization.

Micro-Hydro Manufacturing
How to make small hydro units from car alternators.

Mitigation
Action to reduce sources or enhance sinks of greenhouse gases.

Module
A PV module is composed of PV cells that are encapsulated (typically between two glass covers or between a glass front and a and a suitably weatherproof backing) to protect them against the elements. Such modules, usually framed, are the basic building blocks of PV power systems.

Mushroom Growing
Ways to quickly improve nutrition. Medicinal benefits and mushrooms as a significant tool for soil restoration, replenishment and remediation.

Natural Gas
A mixture of gaseous hydrocarbons, primarily methane, occurring naturally in the earth and used principally as a fuel.

Natural Gas Distribution System
This term generally applies to mains, services, and equipment that carry or control the supply of natural gas from a point of local supply, up to and including the sales meter.

Natural Gas Transmission System
Pipelines installed for the purpose of transmitting natural gas from a source or sources of supply to one or more distribution centres.

Near Neat Fuel
Fuel that is virtually free from admixture or dilution.

Neat Alcohol Fuel
Straight or 100% alcohol (not blended with gasoline), usually in the form of either ethanol or methanol.

Neat Fuel
Fuel that is free from admixture or dilution with other fuels.

No regrets policy
Policies that would generate net social benefits whether or not there is climate change; for example, the value of reduced energy costs or local pollution may exceed the costs of cutting the associated emissions.

Non-Road Vehicle (off-road vehicle)
A vehicle that does not travel streets, roads, or highways. Such vehicles include construction vehicles, locomotives, forklifts, tractors, golf carts, and so forth.

OEM
Original Equipment Manufacturer.

Octane Enhancer
Any substance such as MTBE, ETBE, toluene and xylene that is added to gasoline to increase octane and reduce engine knock.

Octane Rating (Octane Number)
A measure of a fuel's resistance to self ignition, hence a measure as well of the antiknock properties of the fuel.

Operating margin
The operating margin refers to the changes in the operation of plants in an existing power system in response to a project-based activity (e.g. CDM).

Original Equipment Manufacturer (OEM)
The original manufacturer of a vehicle or engine.

Oxides of Nitrogen (NO_x)
Regulated air pollutants, primarily NO and NO_2 but including other substances in minute concentrations. Under the high pressure and temperature conditions in an engine, nitrogen and oxygen atoms in the air react to form various NO_x. Like hydrocarbons, NO_x are precursors to the formation of smog. They also contribute to the formation of acid rain.

Oxygenate
A term used in the petroleum industry to denote fuel additives containing hydrogen, carbon and oxygen in their molecular structure. Includes ethers such as MTBE and ETBE and alcohols such as ethanol and methanol.

Oxygenated Fuels
Fuels blended with an additive, usually methyl tertiary butyl ether (MTBE) or ethanol to increase oxygen content, allowing more thorough combustion for reduced carbon monoxide emissions.

Oxygenated Gasoline
Gasoline containing an oxygenate such as ethanol or MTBE. The increased oxygen content promotes more complete combustion, thereby reducing tailpipe emissions of CO.

Ozone
Tropospheric ozone (smog) is formed when volatile organic compounds (VOCs), oxygen, and NOx react in the presence of sunlight(not to be confused with stratospheric ozone, which is found in the upper atmosphere and protects the earth from the sun's ultraviolet rays). Though beneficial in the upper atmosphere, at ground level, ozone is a respiratory irritant and considered a pollutant.

Paraffins
Group of saturated aliphatic hydrocarbons, including methane, ethane, propane and butane and noted by the suffixane.

Particulate Trap
Diesel vehicle emission control device that traps and incinerates diesel particulate emissions after they are exhausted from the engine but before they are expelled into the atmosphere.

Peak watt (Wp)
The peak watt is the unit by which the power output of PV modules is rated. It is defined as output under peak sunshine conditions of 25°C and irradiance of 1 kilowatt per square meter ($1kW/m^2$ - roughly equivalent to the energy provided by the sun at noon in summer). Thus, if a 50Wp module is mounted in the topics, at midday it should generate about 50W of electrical power; in practice, of course, the output will be slightly less, as the module will likely be operating at a temperature above 25°, which reduces conversion efficiency.

Petroleum Fuel
Gasoline and diesel fuel.

Permanence
The longevity of a carbon pool and the stability of its stocks, given the management and disturbance environment in which it occurs.

Phase Separation
The phenomenon of a separation of a liquid or vapour into two or more physically distinct and mechanically separable portions or layers.

Photography, Audio-Video Production
Training people to earn an income by taking pictures, recording video and sound.

Photovoltaics
The use of lenses or mirrors to concentrate direct solar radiation onto small areas of solar cells, or the use of flat-plate photovoltaic modules using large arrays of solar cells to convert the sun's radiation into electricity.

Policies and measures
Action by government to promote emissions reductions by businesses, individuals, and other groupings; measures include technologies, processes, and practices; policies

include carbon or other energy taxes and standardized fuel-efficiency standards for automobiles.

Portable Fuelling System
A system designed to deliver natural gas to fuelling stations. Such systems are usually configured as tube trailers and are mobile. Fuel delivery usually occurs via over-the-road vehicles.

Practice
An action or set of actions that affect the land, the stocks of pools associated with it or otherwise affect the exchange of greenhouse gases with the atmosphere.

Precision
The repeatability of a measurement (e.g., the standard error of the sample mean).

Private Fleet
A fleet of vehicles owned by a non-government entity.

Propane (C_3H_8)
A gas whose molecules are composed of three carbon and eight hydrogen atoms. Propane is often present in natural gas, and is also refined from crude petroleum. Propane contains about 18 kWh per cubicmeter. In liquefied form called LPG.

Public Fuelling Station
Refers to fuelling station that is accessible to the general public.

Pump Octane
The octane as posted on retail gasoline dispensers as (R+M)/2; same as Antiknock Index.

Rainwater Harvesting
How to collect and store rain; the tools and techniques.

Reformulated Gasoline (RFG)
Gasolines that have had their compositions and/or characteristics altered to reduce vehicular emissions of pollutants.

Refuelling Emissions
VOC vapours that escape from the vehicle fuel tank during refuelling.

Regeneration
The renewal of a stand of trees through either natural means (seeded on-site or adjacent stands or deposited by wind, birds, or animals) or artificial means (by planting seedlings or direct seeding).

Renewables
Energy sources that, within a time frame that is brief relative to the earth's natural cycles, are sustainable; examples are non-carbon technologies such as solar energy, hydropower, waves and wind, as well as carbon-neutral technologies such as biomass.

Reid Vapour Pressure (RVP)
A standard measurement of a liquid's vapour pressure in psi at 37,8 degrees celsius. It is an indication of the propensity of the liquid to evaporate.

Research Octane Number (RON)
The octane as tested in a single-cylinder octane test engine operated under less severe operating conditions. RON affects low to medium-speed knock and engine run-on. Research Octane is presented by the designation R in the (R+M)/2 equation and is the higher of the two numbers.

Retrofit
To change a vehicle or engine after its original purchase, usually by adding equipment such as conversion systems.

Sanitation
How to improve waste water treatment, make composting toilets, grey water systems; from small households to large villages.

Sequestration
The process of removing and storing carbon dioxide from the atmosphere through, for example, land-use change, afforestation, reforestation, or enhancements of carbon in agricultural soils.

Shelter
Improved building methods and materials, using cob, straw, adobe, passive solar design, etc.; the best of the new cutting-edge architectural breakthroughs.

Shifting Agriculture
A form of forest use common in tropic forests where an area of forest is cleared, or partially cleared, and used for cropping for a few years until the forest regenerates. Also known as "slash and burn agriculture," "moving agriculture," or "swidden agriculture."

Sink
Any process or mechanism that removes a greenhouse gas from the atmosphere. A given pool (reservoir) can be a sink for atmospheric carbon if, during a given time interval, more carbon is flowing into it than is flowing out.

Sinks
Places where CO_2 can be absorbed – the oceans, soil and detritus and land biota (trees and vegetation).

Smog
A visible haze caused primarily by particulate matter and ozone. Ozone is formed by the reaction of hydrocarbons and NOx in the atmosphere.

Soil Carbon Pool
Used here to refer to the relevant carbon in the soil. It includes various forms of soil organic carbon (humus) and inorganic soil carbon and charcoal. It excludes soil biomass (e.g., roots, bulbs, etc.) as well as the soil fauna (animals).

Solar Cell
The cell is the component of a photovoltaic system that converts light to electricity. Certain materials (silicon is the most used) produce a PV *effect* in reaction to sunlight; that is, sunlight frees electrons from sites within the materials. The solar cells are structured so that these freed electrons cannot return to their positively charged sites (or "holes") without flowing through an external circuit, thus generating a current. The cells are designed to absorb as much light as possible, but their individual outputs are small; hence, they are interconnected to provide power at an appropriate voltage.

Solar Cooking
How to use the sun to cook, make cookers for resale and/or micro-enterprise, from small household cookers to systems capable of cooking over 1000 meals per day; 2 weeks.

Solar Electric Training and Demonstration
Teaching people how to design, size, install, and maintain photovoltaic systems.

Solar Greenhouses
Growing food during cold months without a fuel source; passive solar greenhouse construction and use.

Solar Pasteurising
Using the sun to purify water, create a mini-mfg/assembly business (cost EUR 30 for family of 4).

Solar Space Heating
Solar heating methods that can be made or assembled locally.

Source
Opposite of sink. A carbon pool (reservoir) can be a source of carbon to the atmosphere if less carbon is flowing into it than is flowing out of it.

Spark Ignition Engine
Internal combustion engine, also called Otto motor, in which the charge is ignited electrically (e.g., with a spark plug).

Spill-over effect
The economic effects of domestic or sectoral mitigation measures on other countries or sectors, which can be positive or negative and include effects on trade, carbon leakage, and the transfer and diffusion of environmentally sound technologies.

Stakeholders
People or entities with interests that would be affected by a particular action or policy.

Stand
See "Forest Stand."

Standardised baseline emission rate
A standardised (or multi-project) baseline emission rate can be calculated without reference to an individual project, based on a pre-defined methodology and characteristics of the regional power system.

Subsidy
A direct payment from the government to an entity, or a tax reduction to that entity, for implementing a practice the government wishes to encourage; greenhouse gas emissions can be discouraged by reducing fossil-fuel subsidies or granting subsidies for insulating buildings or planting trees.

Sulphur Dioxide (SO_2)
An EPA criteria pollutant.

Suspended Particles
Solid particles carried into the atmosphere with the gaseous products of combustion.

Tailpipe Emissions
EPA-regulated vehicle exhaust emissions released through the vehicle tailpipe. Tailpipe emissions do not include evaporative and refuelling emissions.

Tax Incentives
In general, a means of employing the tax code to stimulate investment in or development of a socially desirable economic objective without direct expenditure from the budget of a given unit of government. Such incentives can take the form of tax exemptions or credits.

Technology transfer
An exchange of knowledge, money, or goods that promotes the spread of technologies for adapting to or mitigating climate change; the term generally refers to the diffusion of technologies and technological co-operation across and within countries.

Tertiary Amyl Ethyl Ether (TAEE)
An ether based on reactive C5 olefins and ethanol.

Tertiary Amyl Methyl Ether (TAME)
An ether based on reactive C5 olefins and methanol.

Tetraethyl Lead or Lead
An octane enhancer. One gram of lead increases the octane of one gallon of gasoline about 6 numbers.

Toluene
Basic aromatic compound derived from petroleum and used to increase octane. The most common hydrocarbon purchased for use in increasing octane.

Toxic Emission
Any pollutant emitted from a source that can negatively affect human health or the environment.

Toxic Substance
A generic term referring to a harmful substance or group of substances. Typically, these substances are especially harmful to health. Technically, any compound that has the potential to produce adverse health effects is considered a toxic substance.

Transesterification
A process in which organically-derived oils or fats are combined with alcohol (ethanol or methanol) in the presence of a catalyst to form esters (ethyl or methyl ester).

Transportation Control Measures (TCM)
Restrictions imposed by state or local governments to limit use or access by vehicles during certain times or subject to specific operating requirements, e.g., high-occupancy vehicle lanes.

Tropospheric Oxone (O_3)
Oxygen in condensation form in the lowest stratum of the atmosphere, otherwise known as smog.

Uptake
The addition of carbon to a pool. A similar term is "sequestration".

Vapour Pressure or Volatility
The tendency of a liquid to pass into the vapour state at a given temperature. With automotive fuels, volatility is determined by measuring RVP.

Variable Fuel Vehicle (VFV)
A vehicle that has the capacity of burning any combination of gasoline and an alternative fuel. Also known as a flexible fuel vehicle.

Vehicle Conversion
Retrofitting a vehicle engine to run on an alternative fuel.

Volatile Organic Compound (VOC)
Reactive gas released during combustion or evaporation of fuel. VOCs react with NOx in the presence of sunlight and form ozone.

Voluntary measures
Measures to reduce greenhouse gas emissions that are adopted by firms or other actors in the absence of government mandates; they can involve making climate-friendly products or processes more readily available or encouraging to incorporate environmental values in their market choices.

Water Pumping
How to size and specify the correct pump, how to install and maintain systems.

Water Pump Manufacturing
How to build and maintain "ram" water pumps that use the force of water falling to pump a percentage of the water uphill; making manually operated well pumps.

Watershed Management
Protecting, preserving, and restoring one of our most essential resources.

Wood Alcohol
See Methanol.

Wood Products
Products derived from the harvested wood from a forest, including fuel wood and logs and the products derived from them such as sawn timber, plywood, wood pulp, paper, etc.

Xylene
An aromatic hydrocarbon derived from petroleum and used to increase octane. Highly valued as a petrochemical feedstock. Xylene is highly photochemically reactive and, as a constituent of tailpipe emissions, is a contributor to smog formation.

Zero Emission Vehicle (ZEV)
A vehicle that emits no tailpipe exhaust emissions. ZEV credits can be banked within the Consolidated Metropolitan Statistical Area.

2 Abbreviations and Acronyms

☐	unknown or zero
~	Approximately
<	less than
>	greater than
10^{12}	tera (T)
10^{15}	peta (P)
10^{18}	exa (E)
10^{3}	kilo (k)
10^{6}	mega (M)
10^{9}	giga (G)
AC	Alternating current
API	American Petroleum Institute
B/d	Barrels/day
bbl	barrel
bcm	billion cubic metres
billion	10^{9}
BOO	build, own, operate
BOT	build, operate, transfer
bscf	billion standard cubic feet
Btu	British thermal unit
BWE	Bundesverband Wind-Energie
CEEC	Central and Eastern European Countries
CH_4	Methane
CHP	Combined production of heat and power
CIS	Commonwealth of Independent States
cm	centimetre
CNG	compressed natural gas
CO	Carbon monoxide
CO_2	Carbon dioxide
Convention	United Nations Framework Convention on Climate Change
CUM or cum	Cubic metres
d	day
DC	direct current
DOWA	deep ocean water applications
ECE	Economic Commission for Europe
EJ	Exajoule
EREF	European Renewable Energy Federation
ETBE	ethyl tertiary butyl ether
EU	European Union
EUROSOLAR	European Association for Renewable Energies
FAO	UN Food and Agriculture Organisation
FBC	Fluidised Bed Combustion
FC	Fuel cell
FSU	Former Soviet Union

GCV	Gross calorific value
GDP	Gross domestic product
GEF	Global Environment Facility
GHG	greenhouse gas
GJ	Gigajoule (10^9 joules)
GNP	Gross national product
GOs	Governmental organisations
GTCC	Gas turbine/steam turbine combined cycle system
Gtoe	Giga tonnes of oil equivalent
GW	Gigawatt (10^9 watts)
GW_{el}	gigawatt electricity
GWh	Gigawatt hour
GWP	Global warming potential
h	hour
ha	hectare
HC	Hydrocarbon
HWR	heavy water reactor
Hz	hertz
IAEA	International Atomic Energy Agency
IBRD	International Bank for Reconstruction and Development
IEA	International Energy Agency
IIASA	International Institute for Applied Systems Analysis
IPCC	Intergovernmental Panel on Climate Change
IPP	Independent power producer
IRENA	International Renewable Energy Agency
J	Joule
kboe	Thousand barrels of oil equivalent
kcal	Kilocalorie
kg	Kilogram
kgcr	Kilogram of coal replacement
kgoe	Kilogram of oil equivalent
kl	Kilo litres (thousand litres)
km	Kilometre
Km^2	Square kilometre
kt	Thousand tonnes (10^3 metric tonnes)
ktoe	Thousand tonnes (10^3 metric tonnes) of oil equivalent
KW_{el}	Kilowatt electricity
kWh	Kilowatt-hour (10^3 watts per one hour)
kW_p	kilowatt peak
kW_t	kilowatt thermal
lb	pound (weight)
LNG	Liquefied natural gas
LPG	Liquefied petroleum gas
m	Metre
m/s	metres per second
m^2	square metre
m^3	cubic metre

mb	Millibar
MJ	Megajoule
Ml	Megalitre
mm	Millimetre
MNES	Ministry of Non-conventional Energy sources (India)
mPa s	millipascal second
MPa	Megapascal
mt	million tonnes
Mt	Million tonnes (10^6 metric tonnes)
MtC	Million tonnes of carbon
MtCe	Million tonnes of carbon equivalent
Mtce	Million tonnes of coal equivalent
Mtcr	Million tonnes of coal replacement
Mtoe	Million tonnes of oil equivalent
MW	Megawatts (10^6 watts)
MW_{el}	Megawatt electricity
MWh	Megawatt hour (10^6 watts x one hour)
MW_p	Megawatt peak
MW_t	Megawatt thermal
N	Negligible
N_2O	Nitrous oxide
NCV	Net calorific value
NEA	Nuclear Energy Agency
NGL's	natural gas liquids
NGOs	Non-governmental organisations
Nm^3	normal cubic metre
NO_x	Nitrogen oxides (or oxides of nitrogen)
NPP	Net primary production
NPP	nuclear power plant
OAPEC	Organisation of Arab Petroleum Exporting Countries
OECD	Organisation for Economic Co-operation and Development
OLADE	Latin American Energy Organisation
OPEC	Organisation of the Petroleum Exporting Countries
OTEC	Ocean thermal energy conversion
OWC	Oscillating water column
p.a. or pa	Per annum
Parties	Parties to the UN Framework Convention on Climate Change
PJ	Petajoule (10^{15} joules)
ppm	Parts per million
PPP	Purchasing Power Parity
PV	Photovoltaic
R&D	Research and Development
RD&D	research, development and demonstration
R/P	reserves/production
RAP	Regional Office for Asia and the Pacific
rpm	revolutions per minute
SER	Survey of Energy Resources

SO_2	Sulphur dioxide
t	tonne (metric ton)
tC	Tonne of carbon ($^{44}/12$ tonne CO_2)
tce	tonne of coal equivalent
tcf	trillion cubic feet
TFC	Total final consumption of energy (IEA definition)
TJ	Terajoule
toe	Tonne of oil equivalent (10^7 kcal)
tpa	tonnes per annum
TPES	Total primary energy supply (IEA definition)
trillion	10^{12}
ttoe	thousand tonnes of oil equivalent
tU	tonnes of uranium
TWh	Terawatt hour (10^{12} per one hour)
U	Uranium
UN	United Nations
UNDP	United Nations Development Programme
UNFCCC	United Nations Framework Convention on Climate Change
VAT	Value added tax
W	Watt
WB	World Bank
WCRE	World Council for Renewable Energies
WEC	World Energy Council
WEO	World Energy Outlook
WWEA	World Wind Energy Association
W_p	watts peak
WRI	World Resource Institute
wt	Weight
yr	Year

3   Conversion Factors

Units and Conversions

General conversion factors for energy:

To:	TJ	Gcal	Mtoe	MBtu	GWh
From:	*multiply by:*				
TJ	1	238.8	2.388×10^{-5}	947.8	0.2778
Gcal	4.1868×10^{-3}	1	10^{-7}	3.968	1.163×10^{-3}
Mtoe	4.1868×10^{4}	10^{7}	1	3.968×10^{7}	11630
MBtu	1.0551×10^{-3}	0.252	2.52×10^{-8}	1	2.931×10^{-4}
GWh	3.6	860	8.6×10^{-5}	3412	1

Convertion factors for mass:

To:	kg	t	lt	st	lb
From:	*multiply by:*				
kilogram (kg)	1	0.001	9.84×10^{-4}	1.102×10^{-3}	2.2046
tonne (t)	1000	1	0.984	1.1023	2204.6
long ton (lt)	1016	1.016	1	1.12	2240
short ton (st)	907.2	0.9072	0.893	1	2000
pound (lb)	0.454	4.54×10^{-4}	4.46×10^{-4}	5.0×10^{-4}	1

Convertion factors for volume:

To:	gal U.S.	gal U.K.	bbl	ft³	l	m³
From:	*multiply by:*					
U.S. gallon (gal)	1	0.8327	0.02381	0.1337	3.785	0.0038
U.K. gallon (gal)	1.201	1	0.02859	0.1605	4.546	0.0045
barrel (bbl)	42	34.97	1	5.615	159	0.159
cubic foot (ft³)	7.48	6.229	0.1781	1	28.3	0.0283
litre	0.2642	0.220	0.0063	0.0353	1	0.001
cubic metre (m³)	264.2	220.0	6.289	35.3147	1000	1

Decimal prefixes:

10^1	deca (da)	10^{-1}	deci (d)	
10^2	hecto (h)	10^{-2}	centi (c)	
10^3	kilo (k)	10^{-3}	milli (m)	
10^6	mega (M)	10^{-6}	micro (μ)	
10^9	giga (G)	10^{-9}	nana (n)	
10^{12}	tera (T)	10^{-12}	pico (p)	
10^{15}	peta (P)	10^{-15}	femto (f)	
10^{18}	exa (E)	10^{-18}	atto (a)	

Here are some of the conversion factors you may need to assess your site's feasibility:

1 cubic foot (cf)	7.48 gallons;
1 cubic foot per second (cfs)	448.8 gallons per minute (gpm);
1 inch	2.54 centimeters;
1 foot	0.3048 meters;
1 meter	3.28 feet;
1 cf	0.028 cubic meters (cm);
1 cm	35.3 cf;
1 U.S. gallon	3.785 litres;
1 cuff	28.31 litres;
1 cfs	1,698.7 litres per minute;
1 cubic meter per second (cm/s)	15,842 gpm;
1 pound per square inch (psi) of pressure	2.31 feet (head) of water;
1 pound (lb)	0.454 kilograms (kg);
1 kg	2.205 lbs;
1 kilowatt (kW)	1.34 horsepower (hp);
1 hp	746 Watts.

Energy Conversion and the Related WEC Conversions

In this Statement the conversion convention is the same as that used in ETW, namely that the generation of electricity from hydro (large and small scale), nuclear, and other new renewables (wind, solar, geothermal, oceanic but excluding modern biomass), has a theoretical efficiency of 38.46%. This convention, together with the use of the actual efficiencies (based on the low heating value) for plants using oil or oil products, natural gas or solid fuels (coal, lignite and biomass), guarantees a good comparability in terms of primary energy. However, for the record, WEC has now adopted in all of its recent publications the new conversion convention used by the IEA. New renewables and hydro are assumed to have a 100% efficiency conversion, except for geothermal (10% efficiency). For nuclear plants (excluding breeders) the theoretical efficiency is 33%. For the sake of continuity with ETW, these new conventions are <u>not</u> used in this Statement.

Conversion Factors and Energy Equivalents

1	calorie (cal)	=	4.18	J
1	joule (J)	=	0.239	cal
1000	KWh	=	3.6	GJ
1	tonne of oil equivalent (net, low heat value)	=	42	GJ = 1 toe
1	tonne of coal equivalent (standard, LHV)	=	29.3	GJ = 1 tce
1000	m^3 of natural gas (standard, LHV)	=	36	GJ
1	tonne of natural gas liquids	=	46	GJ
1	toe	=	10 034	Mcal
1	tce	=	7000	Mcal
1000	m^3 of natural gas	=	8600	Mcal
1	tonne of natural gas liquids	=	1000	Mcal
1	tce	=	0.697	toe
1000	m^3 of natural gas	=	0.857	toe
1	tonne natural gas liquids	=	1.096	toe
1000	kWh	=	0.086	toe
1	tonne of fuelwood	=	0.380	toe
1	barrel of oil	=	159	litres
1	barrel of oil	=	Approx. 0.136	tonnes
1	cubic foot	=	0.0283	cubic metres

Because of rounding, some totals may not exactly equal the sum of their component parts, and some percentages may not agree exactly with those calculated from the rounded figures used in the tables.

Conversion Factors and Energy Equivalents

Basic Energy Units

1 joule (J)	=	0.2388 cal		
1 calorie (cal)	=	4.1868 J		
(1 British thermal unit [Btu]	=	1.055 kJ	=	0.252 kcal)

WEC Standard Energy Units

1 tonne of oil equivalent (toe)	=	42 GJ (net calorific value)	=	10 034 Mcal
1 tonne of coal equivalent (tce)	=	29.3 GJ (net calorific value)	=	7 000 Mcal

Note: the tonne of oil equivalent currently employed by the International Energy Agency and the United Nations Statistics Division is defined as 10^7 kilocalories, net calorific value (equivalent to 41.868 GJ)

Volumetric Equivalents

1 barrel	=	42 US gallons	=	approx. 159 litres
1 cubic metre	=	35.315 cubic feet	=	6.2898 barrels

Electricity

1 kWh of electricity output	=	3.6 MJ	=	approx. 860 kcal

Representative Average Conversion Factors

1 tonne of crude oil	=	approx. 7.3 barrels
1 tonne of natural gas liquids	=	45 GJ (net calorific value)
1 000 standard cubic metres of natural gas	=	36 GJ (net calorific value)
1 tonne of uranium (light-water reactors, open cycle)	=	10 000 – 16 000 toe
1 tonne of peat	=	0.2275 toe
1 tonne of fuelwood	=	0.3215 toe
1 kWh (primary energy equivalent)	=	9.36 MJ = approx. 2 236 Mcal

Note: actual values vary by country and over time

Because of rounding, some totals may not agree exactly with the sum of their component parts

4   Inventory of PV systems for sustainable agriculture and rural development

Type of PV Application	Typical System Design	Existing Examples
Applications in the agricultural sector		
Lighting and cooling for poultry factory for extended lighting and increased production	50-150 Wp, electronics, battery, several TL-lights, fan	Egypt, India, Indonesia, Vietnam, Honduras
Irrigation	900 Wp, electronics, small DC or AC pump and water tank	India, Mexico, Chile
Electric fencing for grazing management	2-50 Wp panel, battery fence charger	USA, Australia, New Zealand, Mexico, Cuba
Pest control (moth)	Solar Lanterns used to attract moths away from field	India (Winrock Intl.)
Cooling for fruit preservation	PV/wind hybrid systems or 300-700 Wp PV with DC refrigerators (up to 300 lt.)	Indonesia (Winrock Intl.)
Veterinary clinics	300 Wp, batteries, electronics refrigerator/freezer, 2 TL-lights	Syria (FAO project)
Cattle watering	900 Wp, electronics DC/AC pump, water reservoir	USA, Mexico, Australia
Aeration pumps for fish and shrimp farms	800 Wp, batteries (500 Ah), electronics, DC engine, paddle wheel, for 150m² pond	Israel, USA
Egg incubator	Panel up to 75 Wp, integrated box + heating element for hatching 60 eggs	India (Tata/BPSolar), Philippines (BIG-SOL project)
Crop spraying	5 Wp, sprayer	India (southern states), but cancelled from product package by BPSolar

Type of PV Application	Typical System Design	Existing Examples
Applications in cottage industry		
Tailor workshop	50-100 Wp system with DC lights and electric sewing machine	Several countries (i.e. NREL projects)
Electronic repair workshop	50-100 Wp for DC lights and soldering iron	Bangladesh (Grameen Shakti project) India, Indonesia
Gold jewellery workshop	60 Wp system with DC lights and soldering iron	Vietnam (SELF project)
Bicicle repair workshop	80 Wp system for DC lights and DC small drill	Conceptual: Vietnam – Ha Tinh Province (IFAD project)
Handicrafts workshop (small woodwork, bamboo, basket weaving, etc.)	60-100 Wp system for DC lights and DC small tools	Nepal, Vietnam
Trekking/eco-tourism lodges	Solar lanterns, SHSs and larger PV systems for lights and refrigeration	Nepal, India, Peru, Trinidad and Tobago
Pearl Farms	0.4 – 1 kW PV system to power craft workshops with drills, pumps, lights & compressor	Examples in French Polynesia (Solar energy)
Applications in the commercial service sector		
Village cinema	100-150 Wp system with DC lights and Colour TV + VCR or satellite	Dominican Republic (ENERSOL project), Vietnam (Solarlab), Honduras
Battery charging stations	0.5 – 3 kWp systems with DC battery chargers for kWh sales to households and micro-enterprices	Morocco (Noor Web), Philippines (NEA), Senegal, Thailand, Vietnam (Solarlab), India, Bangladesh
Micro-utility	50 Wp, electronics, battery, 5 – 7 TL ("rented out")	India, Bangladesh (Grameen Shakti project)
Rent-out of solar lanterns for special occasions (weddings, parties, reunions)	Solar lanterns (5 – 10 Wp)	India (NEC) as part of a youth programme

Type of PV Application	Typical System Design	Existing Examples
Applications in the commercial service sector (continued)		
Lights, radio/TV and small appliances such as blenders for restaurants, shops and bars	20-300 Wp, electronics, battery, appliance, invertor (if necessary)	Many countries, incl. Karaoke bar in Philippines (NEA)
Trekking/eco-tourism lodges	Solar lanterns, SHSs and larger PV systems for lights and refrigeration	Nepal, India, Peru, Trinidad and Tobago, Mexico
Cellular telephone service	A 50 Wp system with 2 lights and a socket to charge cellular phone batteries	Bangladesh (Grameen Shakti project)
Computer equipment in rural offices	8 – 300 Wp systems powering lights, fax, TV, etc.	Bangladesh, Costa Rica, Chile
Internet server for E-commerce	Integrated in multifunctional solar facility (> 1 kW)	West Bank (Greenstar project)
Applications for basic social services		
Health clinics	150-200 Wp, electronics, deep-cycle batteries, small refrigerator/freezer	Many countries (WHO standards)
Potable water pumping	1 –4 kWp, electronics, pump, reservoir (generally no batteries needed)	Many countries, e.g. large project in Sahelian countries (EU-project)
Water purification	PV to power UV or ozone water purifiers (0.2-0.3 Wh/litre)	Many countries, e.g China, Honduras, Mexico, West Bank
Water desalination	1 – 2 kWp needed to power reverse osmosis or other water desalination units for 1 m³ per day	Italy, Japan, USA, Australia, Saudi United Arab Emirates
Internet server for telemedicine	Integrated in multifunctional solar facility (> 1 kW)	West Bank (Greenstar project)
Schools and Training centres	PV systems for powering lights, TV/VCR, PCs	Many countries: China, Honduras, Mexico, the Philippines, Mali
Street light	35/70 Wp, electronics, battery, 1 or 2 CFL	India, Indonesia, the Philippines, Brazil, Mali

Source: FAO-survey and literature review, 1997

5   Solar Restaurant in Chile

The Kitchen with solar cookers...

... and a girl enjoys the solar meal (Danial and Patrice 2003)

6   Biomass Turbo Stove BAFOB 5KW

Multifunctional oven for cooking, baking, grilling, and space heating
(El Bassam and Forstinger, 2003)

7   Illuminating the path

From oil lamp ... *(G. Engleret, FAL)*

... *to solar lamp* *(www.afgan-vait.de)*

... *to solar village* *(ddp)*

8   Renewables and Tourism

Solar thermal facilities produces 270,000 kWh energy, 85% of the hot water needs of the robinson Club Daidalos, Kos, Crete every season

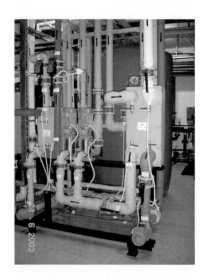

The storage and distribution technology (K. Alektoridis)

Sources of Figures and Tables
(if not stated otherwise)

Figures:
6, 11, 12, 18, 21,22, 25, 28, 31, 32, 33, 34, 35, 36, 39, 40-48, 55, 141, 159, 161, 162, 163: (El Bassam)

16, 17, 19, 20, 23, 24, 26, 27, 29, 30, 37, 38, 49, 50, 51, 56-57, 59, 60-70, 72-77, 79-81, 83-85, 87, 89, 94, 96-102, 104-106, 116, 121, 122-124, 126, 135-137, 150, 160, 169, 171-172, 174-175: (Jane Kruse)

149, 151-154: (B.F.Brix)

164-168: (Q. Xi)

All other figures originate from:

www.cmdc.net
www.enn.com
www.enotrak.de
www.erec-renewables.org
www.eren.doe.gov
www.ewea.org
www.fao.org
www.folkecenter.dk
www.fuelcells.org
www.iea.org
www.ifeed.de
www.imakenews.com
www.international.nreca.org
www.itdg.org
www.lm.dk
www.nrel.gov
www.oecd.org
www.ott.doe.gov
www.pyne.co.uk
www.resum.ises.org
www.rolexawards.com
www.rsvp.nrel.gov
www.sfdc.doe.gov
www.SMA.de
www.small-hydro.com
www.solarcooking.org
www.stirling-engine.de
www.sunoven.com
www.thesustainablevillage.com
www.unepti.org
www.unido.org
www.uniseo.org
www.uni-solar.com
www.victroneenerg.com

www.wind-energie.de
www.worldbank.org
www.worldenergy.org
www.worldsolar.de

References

Aguado-Monsonet, M.A. (1998): Importance of the socio-economic impacts of renewable energies for the southern Mediterranean countries, INTERSUDMED Project. Institute for Perspective Technological Studies, Seville.

Aguado-Monsonet, M.A. (1998) Evaluation of the socio-economic impacts of renewable energies, INTERSUDMED Project. Institute for Perspective Technological Studies, Seville

ASAE (1990). Agricultural machinery management data. American Society of Agricultural Engineers

Best, G. (1995) Bioenergy in developing countries, potentials and constraints, Tropenlandwirt, Beiheft 53. Witzenhausen

Bett, A.W.; Keser, S.; Sulima, O.V. (1997) Recent progress in developing of GaSb Photovoltaic Cells; A.F. Loffe Physico-Technical Institute. St. Petersburg, Russia

Beuse, E.; Boldt, J.; Maegaard, P.; Meyer, N.I.; Windeleff, J.; Ostergaard, I. (2000) Vedvarende energi i Danmark. En krønike om 25 opvækstår 1975 – 2000; OVE's Forlag, Aarhus.

BINE-Informationsdienst (1999) Energie-Datenbank; STN Service Center Europe; FIZ Karlsruhe

Birol, F. (1998) Global energy aspects to 2020, Proc. International Conference on Biomass for Energy and Industry, Würzburg

BMBF (1996) 4. Programm Energieforschung und Energietechnologie

BMU (1996) Sechster Emissionsschutzbericht der Bundesregierung

BWE(2000) "Wind Energy in Europe. Comparison of price regulations with amount regulations", New Energy (1)

Cames, M. et al. (1996) Nachhaltige Energiewirtschaft – Einstieg in die Arbeitswelt von Morgen; Öko-Institut e.V

Chiaramonti, D.; Grimm, P.; Cendagorta, M.; El Bassam, N. (1998) Small agro energy farm scheme implementation for rescuing deserting land in small Mediterranean islands, coastal areas, having water and agricultural land constraints. Feasibility study. In: Proceedings of the 10th International Conference "Biomass for Energy and Industry", Würzburg, 1259-62

CRES (1998) Desalination Guide Using Renewable Energies, Center for Renewable Energy Sources, Greece

Dalsgaard, M.T. (1996) Farm system adjustment in a changing environment. Working paper no.13, Bæredygtige strategier i landbruget

DBU (1997) „ Klimaschutz – eine Investition für die Zukunft", Informationsbroschüre, Osnabrück

De Laquil, P.; Kearney, D.; Geyer, M. and Diver, M. (1993) 'Solar thermal electric technology' , Renewable Energy Sources for Fuels and Electricity, eds Johansonn, T.B.; Kelly, H.; Reddy, A.K.N.; Williams, R.H. and Burham, L., Island Press, Washington D.C.

Deutscher Bundestag (2000) Act on Granting Priority to Renewable Energy Sources, Renewable Energy Sources Act., Berlin

El Bassam, N. (1998) Sustainable development in agriculture – global key issues, Landbauforschung Völkenrode 48(1):1-11

El Bassam, N. (1996) Renewable energy, Potential energy crops for Europe and the Mediterranean region. REU Technical Series 46. Food and Agriculture Organization of the United Nations (FAO)

El Bassam, N. (1998) Biological life support systems under controlled environments. In: El Bassam, N. et. al. (eds.) Sustainable Agriculture for Food, Energy and Industry. James & James Science Publishers, London, Volume 2, 1214-1216

El Bassam, N. (1998) Energy Plant Species – Their Use and Impact on Environment and Development. James & James Science Publishers, London, ISBN 1-873936-70-2

El Bassam, N. (2000) REU-FAO-Project Integrated Energy Farm, FAL/IFEED

El Bassam, N. (2001) Renewable Energy for Rural Communities. Elsevier Science Ltd., Netherlands, Pergamon Renewable Energy 24, 401-408

Ericksen, C. (2002) ISAAC Solar icemaker demonstration http://www.members.aol.com/solariceco

European Commission (1997) Energy technology. ETSU, Oxfordshire, U.K.

EUROSOLAR(1997) Entwicklung und Arbeitsplatzpotential erneuerbarer Energien in der Europäischen Union. In: Solarzeitalter 3

Ferchau, E. (2000) Equipment for Decentralised Cold Pressing of Oil Seeds, Folkecenter for Renewable Energy; Hurup Thy, DK.

Fischedick, M.;. Kaltschmitt, M. (1995) Wind- und Solarstrom im Kraftwerksverbund, C. F. Müller. Heidelberg

Gabler, H.; Zenker, M.; Hein, M. (1997) Modellierung der Energieflüsse in einem Thermophotovoltaik-Generator. Tagungsband Zwölftes Symposium Photovoltaische Solarenergie, Staffelstein, 374-378

Gasch, R. (1996) Windkraftanlagen – Grundlagen und Entwurf; 3. Aufl., Teubner, Stuttgart

Gipe, P. (1999) Wind Energy Basics. A Guide to Small and Micro Wind Systems, Chelsea Green Publishing Company: White River Junction, Vermont, USA and Totnes, England

Global Water Organisation (2001) http://www.globalwater.org/ (Accessed 7/6/2001)

Goetzberger, A. et al. (1997) Sonnenenergie: Photovoltaik; Teubner, Stuttgart

Gombert, A.; Rommel, M. (1997) Breitbandige Antireflexbeschichtung zur Erhöhung der solaren Transmission von Verglasungen an Gebäuden und thermischen Systemen, Tagungsband Forschungsverband Sonnenenergie, Hameln

Greenpeace(1995) Der Preis der Energie – Plädoyer für eine ökologische Steuerreform, Beck'sche Reihe, C.H. Beck, München

Hanus, B. (1997) Das große Anwenderbuch der Windgeneratorentechnik; Franzis, München

Hebling, C.; Wettling, W. (1997) Dünnschicht-Solarzelle aus kristallinem Silizium mit 19% Wirkungsgrad, Erneuerbare Energie, Bd. 2

Heinloth, K. (1996) Energie und Umwelt – Klimaverträgliche Nutzung von Energie, B.G.Teubner, Stuttgart

Heinzel, A.; Nolte, R. (1997) Membranbrennstoffzellen – eine Option für den kleinen Leistungsbereich, Design & Elektronik, München

Hennig, H.-M. (1997) Sorptionsgestützte Klimatisierung mit Solarenergie, Tagungsband Fachseminar „Kälte aus Wärme", Weiterbildungsakademie Weinheim

Ho, G.E ; Harrison, D.G. (1996) Solar Powered Desalination for Remote Areas. Report No. 172. MERIWA project E239. Institute for Environmental Science, Murdoch University, Western Australia

Hoffman, A.R. (2000) Presentation to the World Renewable Energy Conference – VI

IBN – Institut für Baubiologie und Ökologie (1996) Biogasanlagen in der landwirtschaftlichen Wohnen und Gesundheit 79

IEA (2002) : Renewables in global energy supply. IEA Fact Sheet

ISE-Frauhofer Institut Solare Energiesysteme (1997) Jahresbericht Leistungen und Ergebnisse

ISEO (2003) Blueprint for the clean, sustainable energy age, http://www.uniseo.org

IZE, Informationszentrale der Elektrizitätswirtschaft e.V. (1997) Biomasse, Energie, die wächst, Strom Basiswissen Nr.113

Jackson, H.; Svensson, K. (2002) Ecovillage Living: Restoring the Earth and Her People, Green Books Ltd. Totnes, England

Jacobsen, B.H. et al. (1998) An integrated economic and environment farm simulation model (FASSET), Danish Institute of Agricultural and Fisheries Economics, Rapport No. 102

Jansen,W.; Goldworthy, P. (1996) Multidisciplinary research for natural resource management, conceptual and practical implications. Agricultural Systems, Vol.51, pp. 259-279

Kampmann, H. J. (1993) Dieselmotor mit Direkteinspritzung für Pflanzenöl. MTZ Motortechnische Zeitschrift 54

Kerschberger, A. et al. (1998) Transparente Wärmedämmung: Produkte, Projekte, Planungshinweise, Bauverlag, Wiesbaden, Berlin

Kiefer, K.; Hoffmann, V. U. (1997) Erfahrungen mit 2000 Photovoltaik-Dächern in Deutschland. ÖVE-Verbandzeitschrift (10)

Kleemann, M., Meliß, M.(1988) Regenerative Energiequellen, Springer, Berlin, Heidelberg, New York

Kommission der Europäischen Gemeinschaften (1996) Energie für die Zukunft. Erneuerbare Energiequellen - Grünbuch für eine Gemeinschaftsstrategie

Kuratorium für Technik und Bauwesen in der Landwirtschaft (1999) KTBL – Taschenbuch

Langniß, O. et al. (1997) Strategien für eine nachhaltige Energieversorgung – Ein solares Langfristszenario für Deutschland, Deutsches Zentrum für Luft und Raumfahrt DLR, Stuttgart

Ledjeff-Hey, K. et al. (1997) Systemtechnische Aspekte der Membranbrennstoffzellen, Tagungsband GdCh-Jahrestagung, Wien

Leonhardt, W.; Klopfleisch, R.; Jochum, G. (1989) Kommunales Energie–Handbuch, C.F.Müller, Karlsruhe

Lloyd, B.; Lowe, D.; Wilson, L. (2003) Renewable energy in Australian remote communities. Executive summary

Lloyd C.R. (1997)Venco, Reverse Osmosis Unit. NTRC Report #cat 97/5. Centre for Appropriate Technology, N.T., Australia

Luther, J. (1997) Towards an EU-Strategy for Renewable Energies – Research, Development and Demonstration, Proceedings European Congress on Renewable Energy Implementation, Athens, Greece

Maegaard, P. (2003) Wind Energy for the Future, Keynote paper prepared for the 13th Islamic Academy of Science, IAS, Conference, Kuching, Malaysia

Maegaard, P.; Kruse, J. (2001) Energien fra Thy. Fra de lokale til det globale, Nordvestjysk Folkecenter for Vedvarende Energi: Hurup Thy

Marutzky, R. (1997) Moderne Feuerungstechnik zur energetischen Verwertung von Holz und Holzabfällen, Springer VDI, Düsseldorf

Muntwyler, U. (1990) Praxis mit Solarzellen, Kennwerte, Schaltungen und Tips für Anwender, Franzis, München

Myers, (1983) "Energy use for durum wheat production in Tunisia - a case study of 23 farms", Myers, American Society of Agricultural Engineering, Paper 83-3021.

Nielsen, V. (1994) Agricultural production under changed social and environmental conditions, European Union Club of advanced Engineering, Fifth Technical Review, The Netherlands

OECD (1998) Farm management and the environment. OECD Workshop on Agri-environmental indicators, Breakout Session Group 3

Roth, W.; Steinhäuser, A. (1997) Einsatzspektrum von Batterien im Bereich photovoltaisch versorgter Produkte. Design & Elektronik: Entwicklungsforum „Batterien und Ladekonzepte", München

Scheer, H. (2002) The Solar Economy, Renewable Energy for a Sustainable Global Future, Earthscan Publications Ltd. London, Sterling, VA

Scheer, H. (1993) Sonnen-Strategie, Politik ohne Alternative, Piper, München

Scheer, H.; Gandhi, M.; Aitken, D.; Hamakawa, Y.; Palz, W. (1994) Yearbook of Renewable Energies 1994, James & James, London

Seifried, D. (1991) Gute Argumente: Energie, C.H.Beck, München

Stout, B.A. (1990) "Handbook of energy for world agriculture", Elsevier Applied Sciences, London

Tebbutt T.H.Y. (1998) Principles of Water Quality Control. Butterworth-Heinmann, London

Ufheil, M. (1996) Berechnung zur Ermittlung des Raumwärmebedarfs von Gebäuden. Tagungsband Baufachsymposium der Südwestzement, Freiburg

VDI (1996) Erneuerbare Energien. Warum wie sie dringend brauchen aber kaum nutzen. VDI-Verlag

Voss, K. (1997) Transparente Wärmedämmung: Der Stand der Technik. In: Sonnenenergie & Wärmetechnik (1)

Waerras, G. (1998) The socio-economic impact of renewable energy projects in southern mediterranean countries, Technical Report Series / IPTS, Seville

WEC (2003) Energy for people, energy for peace. WEC Statement, WEC Publications, http://www.worldenergy.org

WHO (1996) Water and Sanitation Fact Sheet (112)

Wolf, M. (1998) Erkenntnisse zur Struktur, Synergie und Wirtschaftlichkeit der Energieversorgung aus regenerativen Quellen, 4. Internationale Fachtagung, TU Bergakademie Freiburg

Xi, Q. (1998) Personal communication

Information from the following homepages was also used:

www.afdc.doe.gov
www.bpsolar.com
www.cmdc.net
www.ecosites.net
www.enn.com
www.enotrak.de
www.erec-renewables.org
www.eren.doe.gov
www.ewea.org
www.folkecenter.net
www.gosolar.u-net.com
www.hlf.org.np
www.iea.org
www.ifeed.de
www.imakenews.com
www.international.nreca.org
www.irinnews.org
www.itdg.org
www.nrel.gov
www.nf-2000.org
www.oecd.org
www.ott.doe.gov
www.pyne.co.uk
www.resum.ises.org
www.rolexawards.com
www.rsvp.nrel.gov
www.self.org
www.sfdc.doe.gov
www.SMA.de
www.small-hydro.com
www.solarcooking.org
www.stirling-engine.de
www.sunoven.com
www.thesustainablevillage.com
www.unepti.org
www.unido.org
www.uniseo.org
www.uni-solar.com
www.wcre.org
www.worldbank.org
www.worldenergy.org
www.wwindea.org

Subject Index